Studies in Systems, Decision and Control

Volume 533

The series "Studies in Systems, Decision and Control" (SSDC) covers both new developments and advances, as well as the state of the art, in the various areas of broadly perceived systems, decision making and control–quickly, up to date and with a high quality. The intent is to cover the theory, applications, and perspectives on the state of the art and future developments relevant to systems, decision making, control, complex processes and related areas, as embedded in the fields of engineering, computer science, physics, economics, social and life sciences, as well as the paradigms and methodologies behind them. The series contains monographs, textbooks, lecture notes and edited volumes in systems, decision making and control spanning the areas of Cyber-Physical Systems, Autonomous Systems, Sensor Networks, Control Systems, Energy Systems, Automotive Systems, Biological Systems, Vehicular Networking and Connected Vehicles, Aerospace Systems, Automation, Manufacturing, Smart Grids, Nonlinear Systems, Power Systems, Robotics, Social Systems, Economic Systems and other. Of particular value to both the contributors and the readership are the short publication timeframe and the world-wide distribution and exposure which enable both a wide and rapid dissemination of research output.

Indexed by SCOPUS, DBLP, WTI Frankfurt eG, zbMATH, SCImago.

All books published in the series are submitted for consideration in Web of Science.

Peter S. Sapaty

Spatial Networking in the United Physical, Virtual, and Mental World

Springer

Peter S. Sapaty
Institute of Mathematical Machines
and Systems
National Academy of Sciences of Ukraine
Kyiv, Ukraine

ISSN 2198-4182 ISSN 2198-4190 (electronic)
Studies in Systems, Decision and Control
ISBN 978-3-031-62153-6 ISBN 978-3-031-62154-3 (eBook)
https://doi.org/10.1007/978-3-031-62154-3

This Springer imprint is published by the registered company Springer Nature Switzerland AG
The registered company address is: Gewerbestrasse 11, 6330 Cham, Switzerland

To global consciousness, wisdom, and peace desperately needed to protect the whole world from growing conflicts and wars, which are painfully witnessed especially after working for a long time in very different countries with good cooperation and friends in all of them

Preface

The ideas of this book originate from the WAVE approach which allowed us, more than half a century ago, to implement citywide heterogeneous computer networks and solve distributed problems on them in a simple, flexible, and mobile way, well before the Internet. The invented system organization and management paradigm evolved into Spatial Grasp Technology (SGT) resulting in a European Patent and nine previous books oriented on concrete applications in social and defence systems, security, crises management, gestalt theory, collective robotics, space research, and many others. SGT and its basic Spatial Grasp Language (SGL) allow us to directly exist, move, and operate in physical, virtual, and mental spaces by creating holistic recursive substances self-evolving, self-propagating, and self-matching the distributed worlds. The obtained solutions often exhibit high system integrity, awareness, and even a sort of consciousness.

Many applications and results of the previous works and publications were dealing with graph and network problems and solutions which were extremely important in the researched areas. The current book chooses graphs and networks as the primary and global research objectives after reviewing numerous and very different types and areas of networking. These include transport networks, communication networks, social networks, military networks, economic networks, production networks, evaluation networks, distribution networks, virtual networks, neural networks, psychological networks, criminal networks, etc. We also investigated many existing works on graph and network theory, graph data structures, network and graph algorithms and operations, graph pattern matching, distributed networks, maximum flow in networks, and others, including very important network centrality features.

The main goal of this book is the investigation of applicability of SGT and SGL for the development of higher level infrastructures effectively combining and integrating very different types of networking under the same practical and unified approach, and also the application of the networking approach in very new areas like the creation of artificial organoids and brain research. By the results obtained in the book, the whole world under SGT can be hypothetically and symbolically considered as organized in the following interacting levels: spiritual and mental levels providing global security, integrity, awareness and consciousness, middle networking level materializing from the mental level, and real physical-virtual world controlled and managed by the networking models and cultures.

In our strong belief, the book can help develop advanced social, economic, cultural, educational, and political infrastructures capable of providing and supporting both local and global development, evolution, prosperity, and security within any terrestrial and celestial environments.

Kyiv, Ukraine
April 2024

Peter S. Sapaty

Acknowledgments

Special thanks go to my good colleagues, friends, and their organizations for the lasting support and active discussions of the ideas of this book:

John Page, University of New South Wales, Australia
Steve Lambacher, Aoyama Gakuin University, Tokyo, Japan
Kohei Arai, Saga University, Saga Prefecture, Japan
Bob Nugent, Catholic University of America, Washington D.C.
James White, Vice President of General Magic, retired, Hawaii, USA
Thomas Ditzinger, Editor of Springer books, Germany
Gabriella Williams, Editor, IT—Security, Networking and Communication, Taylor & Francis, UK
Svitlana Tymchyk, Nataliia Karevina, and Yevheniia Petruk, Editorial team of Mathematical Machines and Systems journal, National Academy of Sciences of Ukraine.

Kyiv, Ukraine
June 20, 2024

Peter S. Sapaty

Contents

About the Author

Dr. Peter S. Sapaty Chief Research Scientist, Ukrainian Academy of Sciences, has been with networked systems for more than five decades. Outside of Ukraine, he worked in former Czechoslovakia (now Slovak Republic), Germany, the UK, Canada, and Japan as a group leader, Alexander von Humboldt researcher, and invited and visiting professor. He launched and chaired the SIG on Mobile Cooperative Technologies in Distributed Interactive Simulation project in the US, and invented a distributed control technology that resulted in a European patent and books with Wiley, Springer, Emerald, and Taylor & Francis. He has published more than 280 papers on distributed systems, and has been included in the Marquis Who's Who in the World and Cambridge Outstanding Intellectuals of the 21st Century. Peter also works with several international scientific journals as their editorial board member.

Chapter 1
Introduction

1.1 Networking World and the Global Approach Needed

We are witnessing rapidly growing popularity and activity in the use of networking models and solutions for solving numerous problems in many specific areas. As a universal model and approach, networks can be also used for solving complex problems simultaneously covering very different domains. The main goal of this book is to investigate and propose universal networking techniques and solutions simultaneously covering different areas with united solutions in them, using for this the patented and revealed in many previous publications Spatial Grasp Model and Technology (SGT). The book reviews, analyzes, and classifies the current use of networks in many physical, virtual and mental areas, psychological and psychiatric including. It briefs SGT and its main features, details of Spatial Grasp Language (SGL) and its networked implementation, which can operate without central resources and cover arbitrary large areas. It describes fundamental network operations in SGL which can be performed in parallel and fully distributed mode, including those discovering and analyzing main features of graphs and networks, using active self-matching networking patterns, and providing holistic united solutions in the combined worlds. The book has a special chapter using SGT for modeling organoids and organoid cultures, which are important in medicine and brain research.

The rest of this Introduction chapter is organized as follows. Section 1.2 reviews a number of other books analyzed, compared, used for, and stimulated this book project. Section 1.3 briefs key ideas and features of the Spatial Grasp Model and Technology on which solutions presented in the book are based. Section 1.4 provides summary of other book chapters. References contain literature sources used in the book.

P. S. Sapaty, *Spatial Networking in the United Physical, Virtual, and Mental World*,
Studies in Systems, Decision and Control 533,
https://doi.org/10.1007/978-3-031-62154-3_1

1.2 Other Books Analyzed, Compared, and Used for This Book Project

As this publication was planned as a book, we first looked through some other books published in this and related areas, to learn how they are organized, of which volume, when published, how may relate to the current book proposal, and so on. The following books were investigated in this way and summarized as follows.

The Physical and Virtual Space of the Consulting Room: Room-object Spaces [1]. Examines the role of space and objects in the psychoanalytic process, explores spatialization as simultaneously being a psychological projection of meaning and physically acting upon environment.

Virtual Reality for Physical and Motor Rehabilitation [2]. Reviews two decades of progress in virtual reality for physical and motor rehabilitation, offers research on the capacity of VR to evaluate, address, and reduce motor skill limitations.

The Pilates Method of Physical and Mental Conditioning [3]. A unique system of physical and mental exercise, which stresses control and centering of the body, precise movement, smoothness of motion, proper breathing, and relaxation.

Innovation System Frontiers: Cluster Networks and Global Value [4]. Using interdependencies between key economies analyses systems that cross national borders, shows that technological complexity is an important factor in the formation of production networks.

Foreign Direct Investment and Corporate Networking: A Framework for Spatial Analysis of Investment Conditions [5]. Examines foreign direct investment from spatial perspective and considers how knowledge, regional synergies, economic integration, corporate strategies and networking affect patterns of investment.

Spatial Multidimensional Cooperative Transmission Theories and Key Technologies [6]. Introduces theory and technologies of multi-antenna system, describes spatial multi-dimensional cooperative transmission in the ground-based, air-based and space-based communication systems.

Spatial Network Big Databases: Queries and Storage Methods [7]. Provides a collection of concepts, algorithms, and techniques that effectively harness the power of spatial network big data, investigates scalable graph-based query processing strategies.

Managing Airline Networks: Design, Integration and Innovative Technologies [8]. Discusses the impact of network management on airline resource planning and performance, examines the interplay between network management and adjacent functions.

Optimal Operation of Active Distribution Networks: Congestion Management, Voltage Control and Service Restoration [9]. Provides case studies, modern implementations and supporting flowcharts and code, along with current research in congestion management, service restoration and voltage control of active distribution networks.

Network Psychometrics with R: A Guide for Behavioral and Social Scientists [10]. Provides a comprehensive overview and guide to theoretical foundations of network psychometrics, infers network topology, estimates network parameters from different sources of data.

Centrality Metrics for Complex Network Analysis: Emerging Research and Opportunities [11]. Research findings on centrality metrics and their broader applications for different categories of networks, including wireless sensor networks, curriculum networks, social networks, etc.

Centrality in Strategic Transportation Network Design: An Application to Less-than-truckload Networks [12]. Describes the appearance of a network by transportation network centrality, develops a strategic approach to transportation network design by conceptualizing transportation network centrality.

Crime and Networks [13]. Showcases the use of social networks in the analysis and understanding of various forms of crime, applies to criminology many conceptual and methodological options from social network analysis.

Models of Madness: Psychological, Social and Biological Approaches to Psychosis [14]. Challenges beliefs that madness can be explained without reference to social causes, updates the research showing that hallucinations, delusions etc. are best understood as reactions to adverse life events.

Brain Network Dysfunction in Neuropsychiatric Illness: Methods, Applications, and Implications [15]. Provides a synthesis of the uses of multiple analytic methods applied to neuroimaging data, to seek understanding of the neurobiological bases of psychiatric illnesses.

Human Pluripotent Stem Cell Derived Organoid Models [16]. Highlights recent and emerging advances that describe organoid differentiation protocols for the different organ systems that implement organoids as tools to understand complexity and maturation, high content drug screening, disease modeling, development and evolution.

Brain Organoid Research [17]. Explores multiple methods and approaches used to generate human brain and neuroretinal organoids to address fundamental questions in human brain research.

Organoid Bioengineering—Advances, Applications and Challenges [18]. Organoids are three-dimensional miniature tissue mimics established from embryonic stem cells, human pluripotent stem cells, adult stem cells, and cancer cells. These 3D organoids serve as a valuable tool for fundamental research, disease modeling, drug screening and discovery, and regenerative medicine.

1.3 Basics of the Spatial Grasp Model and Technology

This Spatial Grasp paradigm has been chosen to investigate its applicability as a possible universal approach and technique that could cover very different networking areas and provide important united solutions. Only main technology ideas are summarized below, more can be found in Chaps. 3–5, the related patent and previous books [19–28], latest journal papers [29–36], also previous publications (only few mentioned due to their number and period covered) [37–92].

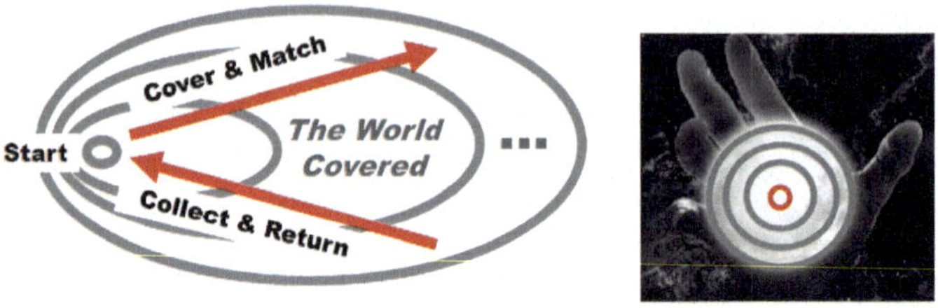

Fig. 1.1 Parallel wavelike world coverage under Spatial Grasp Model

Within Spatial Grasp Model and Technology (SGT), a high-level operational scenario expressed in recursive Spatial Grasp Language (SGL), starting in any world point *propagates, covers, and matches the distributed environment in parallel wave-like mode*, as symbolically shown in Fig. 1.1. Such propagation can result in returning and analyzing the reached states and data, which may be arbitrarily remote, or used for launching more waves.

The SGL allows for organizing *direct space presence and operations* with unlimited power and parallelism. Its universal recursive organization, with operational scenarios called ***grasp***, can be expressed just by a single formula:

$$\boldsymbol{grasp} \rightarrow \boldsymbol{constant}\,|\boldsymbol{variable}|\,\boldsymbol{rule}\ (\{\boldsymbol{grasp},\})$$

The ***rule*** expresses certain action, control, description or context accompanied with operands, which can be any *grasps* too. The rules, starting in certain points, can organize navigation of the world sequentially, in parallel, or any combinations thereof. They can result in the same application points or cause movement to other world points with obtained results left there or returned. The ***SGL interpreter*** consists of specialized modules serving multiple SGL scenarios or their parts happened to be inside this interpreter, also organizing exchanges with other interpreters for distributed SGL solutions. Communicating SGL interpreters can be in arbitrary number of copies, representing ***powerful spatial engines*** *operating without central resources or control*, as in Fig. 1.2. They can effectively work with ***spatial graph and network data*** of any volumes and distributions. Details of SGT are explained in the subsequent Chaps. 3–5.

1.4 Summary of Other Book Chapters

Chapter 2: The Use of Networks in Physical, Virtual, and Mental Domains

The rapidly growing use of networks in various areas of human activity strongly encouraged this book preparation. This chapter reviews, analyzes and classifies many existing publications on the use of networks in very different areas, like

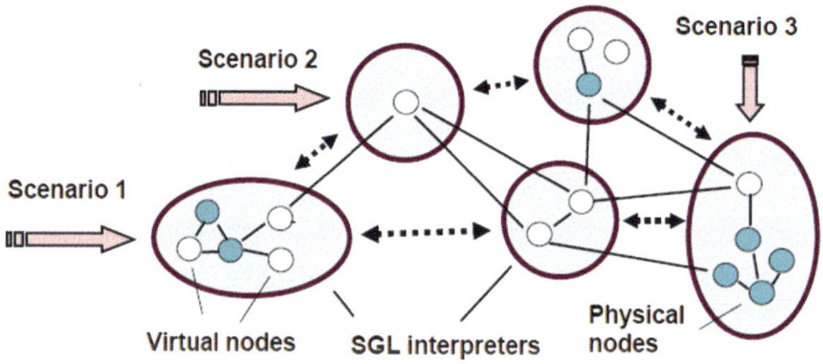

Fig. 1.2 SGL distributed networked interpretation

transport networks, communication networks, social networks, battle and military networks, economic networks, distribution networks, virtual networks, neural networks, psychological networks, and criminal networks. It summarizes existing publications on network analysis and network management too. Then briefs the developed Spatial Grasp Model and Technology (SGT) as a potential global paradigm capable to support, integrate, and manage the reviewed and other heterogeneous networking systems, which is investigated in detail in the subsequent chapters.

Chapter 3: Spatial Grasp Model and Technology Basics

The chapter briefs recursive Spatial Grasp Model and Technology (SGT) which is self-evolving, self-migrating, and self-matching in distributed environments. It describes technology's general issues, Spatial Grasp Language (SGL) features, its top summary, control states, and how SGL scenarios evolve in distributed spaces. Explains details of the networked SGL implementation, SGL interpreter overall organization and its spatial tracking system effectively supporting distributed system integrity, management, and control, altogether allowing SGL interpretation network behave as a universal and powerful spatial engine. Also compares SGT with traditional mobile agents approach based on quite different principles, which were also used by SGT predecessor two decades earlier. Selected references reflect more than half a century publications on this spatial paradigm.

Chapter 4: Spatial Grasp Language (SGL)

The Chapter offers Spatial Grasp Language details which may be particularly useful for better understanding of many practical examples in the subsequent chapters. It first briefs the Spatial Grasp Model and Technology where SGL is its basic language, providing summary of its syntax and main constructs. Then describes different types of SGL constants and variables with their features, spatial distribution and possible movement, also full repertoire of SGL rules with their semantics and practical use, which include type, usage, movement, creation, echoing, verification, assignment, advancement, branching, transference, exchange, timing, qualification, and grasping. Elementary examples of programming in SGL are provided, also references to existing SGL publications and its previous versions called WAVE.

Chapter 5: Organization and Components of SGL Interpreter

The Chapter explains how Spatial Grasp Technology (SGT) can be implemented with communicating Spatial Grasp Language (SGL) interpreter copies embedded into any existing systems, internet including, representing altogether powerful networked engines capable of solving any problems in distributed environments. After briefing SGT and SGL, the Chapter describes main processors and data structures of the interpreter, including its track-based control mechanisms. As both backbone and nerve system, its self-optimizing spatial track system supports hierarchical command and control and remote data and code access, also supervises spatial variables and merges distributed control states and data for decisions and processing at higher organizational levels. With this holistic spatial vision, distributed SGL interpreter can effectively operate without any central control.

Chapter 6: Basic Network Operations in Different Worlds Under SGT

Inheriting the analysis of networking in many areas provided in Chap. 2, this chapter investigates and classifies existing works on main network and graph operations. After briefing the Spatial Grasp Technology and its Spatial Grasp Language (SGL) described in Chaps. 3–5, it presents detailed SGL solutions for network creation, modification, path finding, spanning and shortest path trees, strong and weak components, graph pattern matching, also flows in networks. The latter includes well known Ford Fulkerson method with its solution in SGL which is much simpler and shorter than traditional implementations in Java and C. The networking solutions in SGL can operate on arbitrary large and complex networks, and in highly parallel and fully distributed mode.

Chapter 7: Network Centrality Solutions in SGL

After summarizing the investigated use of networks in many areas of human activity in the previous chapters and briefing the developed Spatial Grasp Technology and its Spatial Grasp Language (SGL), the chapter investigates, evaluates, and models one of the most important features of graphs and networks called Centrality, with its variants Degree Centrality, Closeness Centrality, Betweenness Centrality, and Eigen Vector Centrality. Detailed, efficient, and extremely compact SGL scenarios for solving these centrality problems on a chosen network topology are provided and discussed, being also integral, highly parallel, and fully distributed. They can operate on arbitrary complex and large network structures and evolve in time and space without any central resources.

Chapter 8: Examples of Networking Solutions in Combined Worlds

The chapter investigates applicability of Spatial Grasp Technology (SGT) and its and Spatial Grasp Language (SGL) for development of higher-level social infrastructures effectively integrating very different types of networking. It provides practical examples of combined networking solutions in SGL integrating product development networks expressing design and production of different type of goods, product assessment and request networks describing communities interested in the same or different products, and networks of product delivery to important and registered consumers, with distribution of resultant products via optimal transportation infrastructures. Summary of these combined networking solutions is exhibited too. Deep integration of heterogeneous distributed systems can be organized naturally within global integrity, awareness and consciousness capabilities of SGT.

Chapter 9: Active Spatial Patterns in SGL Versus Traditional Algorithms

Pattern is everything around us; it can represent world's regularity, human-made design, a model, plan or diagram, standard way of modeling, acting and thinking, a distinctive style or form, a combination of qualities and tendencies, etc. The chapter reviews existing works on patterns and pattern languages, considers creating regular patterns and patterns of concrete objects in SGL, offers pattern-based practical solutions for managing transport column, collecting zone coordinates, and tracking mobile objects. It then provides pattern recognition and pattern matching examples in SGL with different types of patterns, explains how SGL can be considered as a real pattern language, concluding with the confirmed effectiveness of SGL and SGT for working with patterns in different areas.

Chapter 10: Networks as Models of Organoid Cultures and Brain Research

This special chapter relates to the concept of organoids, the simplified versions of organs produced artificially, which are of growing importance for disease prediction, prevention, and brain research. Investigates capabilities of creation and evolution of organoid models by Spatial Grasp Technology (SGT) described in the previous chapters. Analyzes and classifies latest organoid-related publications, shows expression in SGL of growing organoid models for cell growth, division, replication, worm movement, killer cell operation, and full organoid creation. It reviews other organoid developments of potential interest to SGT like mini-brains in robotics, organoids with brain waves, and organoids in outer space. Confirms applicability of the developed spatial paradigm to express and simulate cells growth up to the whole organoids.

Chapter 11: Conclusions

The book investigated applicability of Spatial Grasp paradigm for solving networking problems in very different areas. It allows us to directly move and operate in physical, virtual and mental spaces in a clear and compact mode, radically differing from traditional algorithmic thinking but rather creating flexible holistic and intelligent substances self-evolving, self-propagating, self-covering and self-matching the distributed worlds together with solving complex problems in them. By results obtained in the book and discussed in previous publications on this paradigm, the whole world can be symbolically considered as integrating spiritual and mental level providing global awareness and consciousness, middle networking level which materializes from the mental level, and real physical-virtual world controlled from above and inside by networking models.

References

1. Wright, D.L.S.: The Physical and Virtual Space of the Consulting Room: room-object Spaces. Routledge (2022)
2. Weiss, P.L., Keshner, E.A., et al.: Virtual Reality for Physical and Motor Rehabilitation (Virtual Reality Technologies for Health and Clinical Applications). Springer (2014)
3. Friedman, P.: The Pilates Method of Physical and Mental Conditioning. Doubleday (1984)
4. Wixted, B.: Innovation System Frontiers: Cluster Networks and Global Value (Advances in Spatial Science). Springer (2009)
5. Morsink, R.L.A.: Foreign Direct Investment and Corporate Networking: a Framework for Spatial Analysis of Investment Conditions. Edward Elgar Publishing (1999)
6. Bai, L., Liang, X., et al.: Spatial Multidimensional Cooperative Transmission Theories and Key Technologies. World Scientific (2020)
7. Yang, K.S., Shekhar, S.: Spatial Network Big Databases: queries and Storage Methods. Springer (2017)
8. Franke, M.: Managing Airline Networks: design, Integration and Innovative Technologies (Managing Aviation Operations). Routledge (2020)
9. Wu, Q., Shen, F., et al.: Optimal Operation of Active Distribution Networks: congestion Management, Voltage Control and Service Restoration. Academic Press (2023)
10. Isvoranu, A.-M., Epskamp, S., et al.: Network Psychometrics with R: a Guide for Behavioral and Social Scientists. Routledge (2022)
11. Meghanathan, N.: Centrality Metrics for Complex Network Analysis: emerging Research and Opportunities (Advances in Wireless Technologies and Telecommunication). IGI Global (2018)
12. Lange, A.: Centrality in Strategic Transportation Network Design: an Application to Less-Than-Truckload Networks (Edition KWV). Springer Gabler (2019)
13. Morselli, C.: Crime and Networks (Criminology and Justice Studies). Routledge (2013)
14. Read, J., Bentall, R., et al.: Models of Madness: psychological, Social and Biological Approaches to Psychosis (The International Society for Psychological and Social Approaches to Psychosis Book Series). Routledge (2013)
15. Diwadkar, V.A., Eickhoff, S.B.: Brain Network Dysfunction in Neuropsychiatric Illness: methods, Applications, and Implications. Springer (2021)
16. Spence, J.: Human Pluripotent Stem Cell Derived Organoid Models (ISSN Book 159) , 1st edn. Academic Press (2020)
17. Gopalakrishnan, J.: Brain Organoid Research (Neuromethods Book 189), Humana (2022)
18. Paul, M.K.: Organoid Bioengineering—Advances, Applications and Challenges (Biomedical Engineering) IntechOpen (2022)
19. Sapaty, P.S.: A distributed processing system, European Patent N 0389655, Publ. 10.11.93, European Patent Office
20. Sapaty, P.S.: Mobile Processing in Distributed and Open Environments. Wiley, New York (1999)
21. Sapaty, P.S.: Ruling Distributed Dynamic Worlds. Wiley, New York (2005)
22. Sapaty, P.S.: Managing Distributed Dynamic Systems with Spatial Grasp Technology. Springer (2017)
23. Sapaty, P.S.: Holistic Analysis and Management of Distributed Social Systems. Springer (2018)

24. Sapaty, P.S.: Complexity in International Security: a Holistic Spatial Approach. Emerald Publishing (2019)
25. Sapaty, P.S.: Symbiosis of Real and Simulated Worlds under Spatial Grasp Technology. Springer (2021)
26. Sapaty, P.S.: Spatial Grasp as a Model for Space-based Control and Management Systems. CRC Press (2022)
27. Sapaty, P.S.: The Spatial Grasp Model: applications and Investigations of Distributed Dynamic Worlds. Emerald Publishing (2023)
28. Sapaty, P.S.: Providing Integrity, Awareness, and Consciousness in Distributed Dynamic Systems. CRC Press (2024)
29. Sapaty, P.S.: Spatial management of air and missile defense operations. Math. Mach. Syst. (1) (2023)
30. Sapaty, P.S.: Providing distributed system integrity under spatial grasp technology. Math. Mach. Syst. (2) (2023)
31. Sapaty, P.S.: Providing global awareness in distributed dynamic systems. Int. Relat. Dipl. **11**(2), 87–100 (2023). https://doi.org/10.17265/2328-2134/2023.02.002
32. Sapaty, P.S.: Simulating distributed consciousness with spatial grasp model. Math. Mach. Syst. (3) (2023)
33. Sapaty, P.S.: Managing distributed systems with spatial grasp patterns. Math. Mach. Syst. (4) (2023)
34. Sapaty, P.S.: Network centrality operations under spatial grasp technology. J. Adv. Artif. Intell. Mach. Learn. **1**(1), 1–11 (2023)
35. Sapaty, P.S.: Spatial networks as models for organoid cultures and brain research. Math. Mach. Syst. (1) (2024)
36. Sapaty, P.S.: Basic network operations under spatial grasp model. Sci. Set. J. Phys. **3**(1), 01–16 (2024)
37. Bondarenko, A.T., Mikhalevich, S.B., Nikitin, A.I., Sapaty, P.S.:Software of BESM-6 computer for communication with peripheral computers via telephone channels. In: Computer software, vol. 5. Institute of Cybernetics Press, Kiev (1970)
38. Sapaty, P.S.: A method of organization of an intercomputer dialogue in the radial computer systems. In: The Design of Software and Hardware for Automatic Control Systems. Institute of Cybernetics Press, Kiev (1973)
39. Sapaty, S.: A wave language for parallel processing of semantic networks. Comput. Artif. Intell. **5**(4), 289–314 (1986)
40. Sapaty, P.S.: Distributed modeling of cooperative behavior by mobile agents. In: Proceedings of the Sixth Conference on Computer Generated Forces and Behavioral Representation, pp. 599–613. IST UCF, Orlando, FL (1996)
41. Sapaty, P.S.: Distributed artificial brain for collectively behaving mobile robots. In: Proceedings of Symposium and Exhibition Unmanned Systems 2001, Baltimore, MD (2001)
42. Sapaty, P.S.: Crisis management with distributed processing technology. Int. Trans. Syst. Sci. Appl. **1**(1), 81–92 (2006). ISSN 1751-1461
43. Sapaty, P.S.: Global management of distributed EW-related system. In: Proceedings of International Conference Electronic Warfare: operations & Systems 2007, Thistle (2007)
44. Sapaty, P.S.: Distributed technology for global dominance. In: Proceedings of International Conference Defense Transformation and Net-Centric Systems 2008, as part of the SPIE Defense and Security Symposium, World Center Marriott Resort and Convention Center, Orlando, FL, USA (2008); Proceedings of SPIE—Volume 6981, Defense Transformation and Net-Centric Systems 2008, Raja Suresh, Editor, 69810T (2008)
45. Sapaty, P.S.: Distributed technology for global dominance. Keynote lecture. In: Proceedings of the Fifth International Conference in Control, Automation and Robotics ICINCO 2008, The Conference Proceedings, Funchal, Madeira, Portugal (2008)
46. Sapaty, P.S.: Human-robotic teaming: a compromised solution. In: AUVSI's Unmanned Systems North America 2008, San Diego, USA (2008)

47. Sapaty, P.S.: High-level communication protocol for dynamically networked battlefields. In: Proceedings of International Conference Tactical Communications 2009 (Situational Awareness and Operational Effectiveness in the Last Tactical Mile), One Whitehall Place, Whitehall Suite and Reception, London, UK (2009)
48. Sapaty, P.S.: Distributed capability for battlespace dominance. In: Electronic Warfare 2009 Conference and Exhibition, Novotel London West Hotel and Conference Center, London (2009)
49. Sapaty, P.S.: Providing spatial integrity for distributed unmanned systems. In: Proceedings of 6th International Conference in Control, Automation and Robotics ICINCO 2009, Milan, Italy (2009)
50. Sapaty, P.S.: Distributed technology for global control. In: Lecture Notes in Electrical Engineering, Informatics in Control, Automation and Robotics, vol. 37. Springer, Berlin (2009)
51. Sapaty, P.S.: Gestalt-based integrity of distributed networked systems. In: SPIE Europe Security + Defence, BCC Berliner Congress Centre, Berlin, Germany (2009)
52. Sapaty, P.S.: Remote control of open groups of remote sensors. In: Proceedings of SPIE Europe Security + Defence, Berlin, Germany (2009)
53. Sapaty, P.S. (2010) Tactical communications in advanced systems for asymmetric operations. In: Proceedings of Tactical Communications 2010, CCT Venues, Canary Wharf, London UK (2010)
54. Sapaty, P.S.: High-level technology to manage distributed robotized systems. In: Proceedings of Military Robotics 2010, Jolly St. Ermins, London, UK (2010)
55. Sapaty, P.S.: Emerging asymmetric threats, Q&A session. In: Tactical Communications 2010, CCT Venues, Canary Wharf, London UK (2010)
56. Sapaty, P.S.: High-level organization and management of directed energy systems. In: Proceedings of Directed Energy Weapons 2010, CCT, Canary Wharf, London UK (2010)
57. Sapaty, P.S.: Unified transition to robotized armies with spatial grasp technology. In: International Summit Military Robotics, London, United Kingdom (2012)
58. Sapaty, P.S.: Distributed air and missile defense with spatial grasp technology. Intell. Control Autom. Sci. Res. **3**(2), 117–131 (2012)
59. Sapaty, P.S.: Global electronic dominance. In: 12th International Fighter Symposium, Grand Connaught Rooms, London, UK (2012)
60. Sapaty, P.S.: Providing global awareness in distributed dynamic environments. In: International summit ISR, London (2013)
61. Sapaty, P.S.: Ruling distributed dynamic worlds with spatial grasp technology. In: Tutorial at the International Science and Information Conference 2013 (SAI), London, UK (2013)
62. Sapaty, P.S.: Night vision under advanced spatial intelligence: a key to battlefield dominance. In: International Summit Night Vision 2013, London (2013)
63. Sapaty, P.S.: Integration of ISR with advanced command and control for critical mission applications. In: SMi's ISR Conference, Holiday Inn Regents Park, London (2014)
64. Sapaty, P.S.: Unified transition to cooperative unmanned systems under spatial grasp paradigm. Int. J. Trans. Netw. Commun. **2**(2) (2014)
65. Sapaty, P.S.: From manned to smart unmanned systems: a unified transition. In: SMi's Military Robotics, Holiday Inn Regents Park London (2014)
66. Sapaty, P.S.: Unified transition to cooperative unmanned systems under spatial grasp paradigm. In: 19th International Command and Control Research and Technology Symposium, Alexandria, Virginia (2014)
67. Sapaty, P.S.: Distributed human terrain operations for solving national and international problems. Int. Relat. Dipl. **2**(9) (2014)
68. Sapaty, P.S.: Providing over-operability of advanced ISR systems by a high-level networking technology. In: SMI's Airborne ISR, Holiday Inn Kensington Forum, London, United Kingdom (2015)
69. Sapaty, P.S.: Distributed missile defence with spatial grasp technology. In: SMi's Military Space, Holiday Inn Regents Park London (2015)
70. Sapaty, P.S.: Military robotics: latest trends and spatial grasp solutions. Int. J. Adv. Res. Artif. Intell. **4**(4) (2015)

71. Sapaty, P.S.: A brief introduction to the spatial grasp language (SGL). J. Comput. Sci. Syst. Biol. **09**(02) 2015. https://www.researchgate.net/publication/307744279_A_Brief_Introduction_to_the_Spatial_Grasp_Language_SGL
72. Sapaty, P.S.: Organization of advanced ISR systems by high-level networking technology. Math. Mach. Syst. (1) (2016)
73. Sapaty, P.S.: Towards massively robotized systems under spatial grasp technology. J. Comput. Sci. Syst. Biol. **9**(1) (2016)
74. Sapaty, P.S.: Towards wholeness and integrity of distributed dynamic systems. J. Comput. Sci. Syst. Biol. **9**(3) (2016)
75. Sapaty, P.S.: Towards global goal orientation, robustness and integrity of distributed dynamic systems. J. Int. Relat. Dipl. **4**(6) (2016)
76. Sapaty, P.: Spatial grasp language (SGL). Adv. Image Video Process. **4**(1) (2016). https://journals.scholarpublishing.org/index.php/AIVP/article/view/1922/1052
77. Sapaty P.S.: Mosaic warfare: from philosohpy to model to solution. Math. Mach. Syst. (3) (2019)
78. Sapaty, P.S.: Global network management under spatial grasp paradigm. Int. Robot. Autom. J. **6**(3), 134–148 (2020). https://medcraveonline.com/IRATJ/IRATJ-06-00212.pdf
79. Sapaty, P.S.: Global network management under spatial grasp paradigm. Glob. J. Res. Eng.: J. Gener. Eng. **20**(5), 58–81 (Version 1.0) (2020). https://globaljournals.org/GJRE_Volume20/6-Global-Network-Management.pdf
80. Sapaty, P.: Advanced terrestrial and celestial missions under spatial grasp technology. Aeron. Aerosp. Open Access J. **4**(3) (2020). https://medcraveonline.com/AAOAJ/AAOAJ-04-00110.pdf
81. Sapaty, P.: Spatial management of distributed social systems. J. Comput. Sci. Res. **02**(03) (2020). https://ojs.bilpublishing.com/index.php/jcsr/article/view/2077/pdf
82. Sapaty, P.: Towards global nanosystems under high-level networking technology. Acta Sci. Comput. Sci. **2**(8) (2020). https://www.actascientific.com/ASCS/pdf/ASCS-02-0051.pdf
83. Sapaty, P.: Symbiosis of distributed simulation and control under spatial grasp technology, SSRG Int. J. Mob. Comput. Appl. (IJMCA) **7**(2) (2020). http://www.internationaljournalssrg.org/IJMCA/2020/Volume7-Issue2/IJMCA-V7I2P101.pdf
84. Sapaty, P.S.: Symbiosis of real and simulated worlds under global awareness and consciousness. In: Abstract, The Science of Consciousness (TSC) Conferences, Tucson, Arizona (2020). https://eagle.sbs.arizona.edu/sc/report_poster_detail.php?abs=3696
85. Sapaty, P.S.: Spatial grasp as a model for space-based control and management systems. Math. Mach. Syst. (1), 135–138 (2021). http://www.immsp.kiev.ua/publications/articles/2021/2021_1/Sapaty_book_1_2021.pdf
86. Sapaty, P.S.: Managing multiple satellite architectures by spatial grasp technology. Math. Mach. Syst. (1), 3–16 (2021). http://www.immsp.kiev.ua/publications/eng/2021_1/
87. Sapaty, P.S.: Spatial management of large constellations of small satellites. Math. Mach. Syst. (2) (2021). http://www.immsp.kiev.ua/publications/articles/2021/2021_2/02_21_Sapaty.pdf
88. Sapaty, P.S.: Global management of space debris removal under spatial grasp technology. Acta Sci. Comput. Sci. **3**(7) (2021). https://www.actascientific.com/ASCS/pdf/ASCS-03-0135.pdf
89. Sapaty, P.S.: Space debris removal under spatial grasp technology. Netw. Commun. Technol. **6**(1) (2021). https://www.ccsenet.org/journal/index.php/nct/article/view/0/45486
90. Sapaty, P.S.: Relation of spatial grasp paradigm to higher psychological and mental concepts. Acta Sci. Comput. Sci. **4**(12) (2022). https://actascientific.com/ASCS/pdf/ASCS-04-0359.pdf
91. Sapaty, P.S.: Seeing and managing distributed worlds with spatial grasp paradigm. Acta Sci. Comput. Sci. **4**(12) (2022). https://actascientific.com/ASCS/pdf/ASCS-04-0365.pdf
92. Sapaty, P.S.: Comprehending distributed worlds with the spatial grasp paradigm. Math. Mach. Syst. **1** (2022). http://www.immsp.kiev.ua/publications/articles/2022/2022_1/01_22_Sapaty.pdf

Chapter 2
The Use of Networks in Physical, Virtual, and Mental Domains

2.1 Introduction

We are witnessing rapidly growing use of networks and their different types, both physical and virtual, which have become an important part of our lives. The easiest way to expand our network is to build on the relationships with people we know. Networking is beneficial in social and professional aspects, where social networking uses different media platforms to connect with others having similar interests. Networks can also help satisfy objectives or needs, and if something in the transportation network becomes unavailable, their ability to distribute food, clothes and products would be compromised. In today's competitive market, responsiveness to customer or supplier demand is a decisive factor, and the network is considered as one of most critical resources. Similarly, a computer network is transmitting essential data, from one computer to another. The *aim of this chapter* is to review and analyze networking concepts in very different human areas and activities, and select a universal networking model and technology which could be useful for analysis and implementation of global networking.

The rest of this chapter is organized as follows. Section 2.2 describes and reviews different types of networks and their usage which include: transport networks, communication networks, social networks, battle and military networks, economic networks, distribution networks, virtual networks, neural networks, psychological networks, and criminal networks. Section 2.3 reviews existing publications on network analysis, and Sect. 2.4 on network management. Section 2.5 briefs the chosen Spatial Grasp Model and Technology (SGT) described in details in the following sections. Section 2.6 concludes the chapter. References include the used numerous publications and sources on different types of networking and their analysis and management, also existing applications on SGT.

P. S. Sapaty, *Spatial Networking in the United Physical, Virtual, and Mental World*, Studies in Systems, Decision and Control 533,
https://doi.org/10.1007/978-3-031-62154-3_2

2.2 Different Types of Networks and Their Usage

We are providing a brief summary on the examined networks-related publications, explaining their main ideas, orientations, contents, and usage.

2.2.1 Transport Networks

Transport networks [1] include road, rail, air and water transport networks and related infrastructure with links between different networks. Also includes the trans-European transport network as defined in the Decision of the European Parliament and of the Council on Community Guidelines for the development of the trans-European transport network. The transport component should comprise integrated transport network and related features that are seamless within each national border. National transport networks may also be seamless at European level, i.e. connected at national borders. Transportation data includes topographic features related to transport by road, rail, water, and air. It is important that the features form networks where appropriate, and that links between different networks are established, in order to satisfy the requirements for intelligent transport systems such as location based services and telematics.

Transportation networks [2] generally refer to a set of links, nodes, and lines that represent the infrastructure or supply side of the transportation. The links have characteristics such as speed and capacity for roadways; frequency and travel time data are defined on transit links or lines for the transit system. Transportation analysis zones are represented in the network by centroids, a subset of the nodes; and centroid connectors are links that provide access and egress to the system. Transportation networks could be of several types for representing the infrastructure or supply of different modes. Highway and transit networks are a principal means by which the supply side of transportation is represented. Transit networks are a spatial representation of bus, rail, and other types of transit routes available in a region. Links can include fixed-guideway (e.g., passenger rail) to bus lines that operate in mixed-traffic on the roadway links. The network models can be both very precise and accurate.

Multiservice Transport Networks are discussed in [3]. A new generation of SONET and DWDM systems providing the functions of multiple network elements in a single platform has emerged. This new platform is called a Multiservice Provisioning Platform (MSPP). MSPPs are a popular solution for building new networks and upgrading existing networks to take advantage of new services and integration of voice and data. Cisco Systems provides an MSPP product for both service provider and enterprise networks. Cisco Systems is the market leader in MSPP technology in North America. Building Multiservice Transport Networks will become an indispensable reference for Cisco customers and constituents who are deploying MSPP solutions. Building Multiservice Transport Networks teaches all facets of MSPP networks in an easy-to-understand manner and from both the service provider and enterprise perspective. It provides the background material necessary to learn key aspects of SONET, SDH, DWDM, Ethernet, and storage networking, and does so through network diagrams, application examples, design guidelines, and detailed configurations.

Packet Transport Networks are discussed in [4]. Transport networks provide transparent transmission of client data traffic between connected client devices by establishing and maintaining point-to-point or point-to-multipoint connections between such devices. The network is basically independent of any higher-layer network that may exist between clients. In addition to client traffic, a transport network must carry traffic to facilitate its own operation that is necessary for connection control, operation, administration, and maintenance functions, network management systems, and protection, just as traditional dedicated circuit-based transport technologies such as synchronous digital hierarchy and optical transport networks, have satisfactory grade capabilities of these functions.

Air Transport Networks are described in [5]. Air transport network or air transportation network is an example of transport networks and spatial networks. The nodes of the network are the airports and the links represent direct flight routes between two airports. Alternatively, cities can be considered as the nodes with links representing direct flight connection between them. Air transport networks can be defined worldwide as well as for one region or for one airline company; the scale of the network can be global or domestic. The graph of an air transport network is spatial but not a planar graph. The air transportation network is a complex network which has the properties of small-world networks and scale-free networks. The degree distribution of the nodes displays a heavy-tailed distribution. The hubs of the network have large connectivities and long-distance connectivities at the same time.

Geographical routing protocol for vehicular networks is discussed in [6]. The real-time traffic information dissemination among on-road vehicles has been envisioned via realizing vehicular ad hoc networks as smart service-oriented roadside wireless sensor networks. The network enables various types of real-time traffic applications related to safety and infotainment for drivers and passengers. The information dissemination-centric routing protocols for vehicular networks have to dynamically adopt under the constrained network environment while considering the higher mobility of vehicular nodes and unpredictable physical topologies in the network. The issue needs to be addressed through smart and network-aware routing protocols. Geographical routing protocols have witnessed significant attention for information dissemination under these types of dynamic vehicular network environment. The paper presents an improved road segment-based geographical routing protocol focusing on better head node selection for information dissemination. It divides the forwarding area into a number of road segments and selects a head node on each segment by focusing on traffic-aware information including the location, direction, and link quality-centric score for every vehicle on each road segment.

See Fig. 2.1 symbolically related to transport networks.

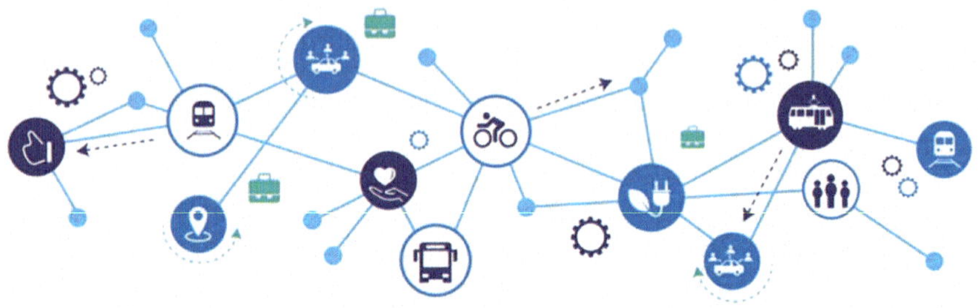

Fig. 2.1 Transport network

2.2.2 Communication Networks

Communication networks are regular patterns of person-to-person relationships through which information flows in an organization. This means that the flow of information is managed, regulated and structured. Communication networks may be formal or informal.

Communication network [7] is a pattern or form that is implemented in the organization to communicate information effectively. The communication network is the established system where the message may flow in one or too many directions in the organization based on requirements. Under the wheel communication network, the leader of the organization works as a central person to accumulate information and disseminate it to all members. The chain network has the appearance of an organizational chain of command. It is a horizontal or sideward form of a circle network where a person can communicate with another person only next to his right or left. In all channel networks, all members can communicate with all the members without any formal restrictions. Communication networks provide efficient flow of information, enhanced collaboration, reduced ambiguity and misunderstanding, increased engagement and employee satisfaction, alignment with organizational goals.

Computing in communication networks is discussed in [8]. It provides comprehensive details and practical implementation tactics on the novel concepts and enabling technologies at the core of the paradigm shift from store and forward to compute and forward in future communication networks and systems. Explains how to create virtualized large scale testbeds using well-established open source software. Shows how and where to place disruptive techniques, such as machine learning, compressed sensing, or network coding in a newly built testbed. In addition, presents a comprehensive overview of current standardization activities. Specific chapters explore upcoming communication networks that support verticals in transportation, industry, construction, agriculture, health care and energy grids, underlying concepts, such as network slicing and mobile edge cloud, enabling technologies, disruptive innovations, compressed sensing and machine learning.

Optimization and control of communication networks is considered in [9]. It provides a modern mathematical approach to the design of communication networks, blending control, optimization, and stochastic network theories. A broad range of performance analysis tools are discussed, including important advanced topics that have been made accessible for the first time. Taking a top-down approach to network protocol design, it begins with the deterministic model and progress to more sophisticated models. Network algorithms and protocols are tied closely to the theory, illustrating the practical engineering applications of each topic. The background behind the mathematical analyses is given before the formal proofs and is supported by worked examples, enabling to understand the big picture before going into the detailed theory. End-of-chapter problems cover a range of difficulties, with complex problems broken into parts and solutions available online.

Introduction to wireless communications and networks is considered in [10]. It provides an intuitive and accessible introduction to the fundamentals of wireless communications and their tremendous impact. Starts with basic information on physics and mathematics and then expands on it, helping to understand fundamental concepts of RF systems and how they are designed. Covering diverse topics in wireless communication systems, including cellular and personal devices, satellite and space communication networks, telecommunication regulation, standardization and safety, it combines theory and practice using problems from industry, and includes examples of day-to-day work in the field. Divided into two parts, the basic and advanced, it includes information on what checks and balances are used by wireless engineers around the globe and addresses questions concerning safety, reliability, and long-term operation.

Underwater communications and networks are considered in [11]. It covers all related communication technologies of underwater wireless communication, such as acoustic communication, optical communication, and magneto-inductive communication. After describing each technology, it relays their pros and cons, as it is essential to learn the underlying mechanism, advancements, and limitations of these techniques. It provides basics fundamentals of the three technologies, their advantages and disadvantages, and their applications, also introduces research trends, pointing in the direction of research in the field of underwater wireless communication. It is essential in the field of underwater communications, also includes end-of-chapter questions and numerical problems.

Network security is considered in [12]. Widely regarded as the most comprehensive yet comprehensible guide to network security and cryptography, the previous editions of Network Security received critical acclaim for lucid and witty explanations of the inner workings of cryptography and network security protocols. This edition has significantly updated and revised the previous content, and added new topics that have become important, explaining sophisticated concepts in a friendly and intuitive manner. For protocol standards, it explains the various constraints and committee decisions that led to the current designs. For cryptographic algorithms, it explains the intuition behind the designs, as well as the types of attacks the algorithms are designed to avoid. It explains implementation techniques that can cause vulnerabilities even if the cryptography itself is sound.

6G mobile wireless networks are discussed in [13]. It is the world's first book on 6G mobile wireless networks that aims to provide a comprehensive understanding of key drivers, use cases, research requirements. It covers 6G use cases, requirements,

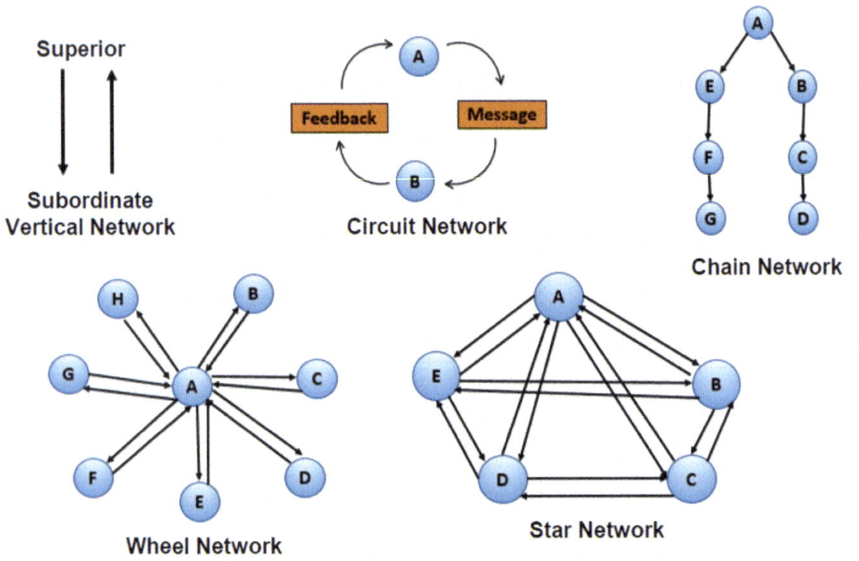

Fig. 2.2 Types of communication networks

metrics and enabling technologies, PHY technologies for 6G wireless, reconfigurable intelligent surface for 6G wireless networks, and many others. The aim of this book is to explore the evolution from current 5G networks towards the future 6G networks from a service, air interface and network perspective, laying out a vision for 6G networks. It provides a systematic treatment of the state-of-the-art in these emerging topics and their role in supporting a wide variety of verticals in the future, also a comprehensive overview of the expected applications of 6G with a detailed discussion of their requirements and possible enabling technologies. Also outlines possible challenges and research directions to facilitate the future research and development of 6G mobile wireless networks.

See Fig. 2.2 symbolically related to communication networks.

2.2.3 Social Networks

Definition of social networkof social network is in [14], which is considered as a social structure made up of a set of social actors (such as individuals or organizations), sets of dyadic ties, and other social interactions between actors. The social network perspective provides a set of methods for analyzing the structure of whole social entities as well as a variety of theories explaining the patterns observed in these structures. The study of these structures uses social network analysis to identify local and global patterns, locate influential entities, and examine network dynamics. Social networks and the analysis of them is an inherently interdisciplinary academic field which emerged from social psychology, sociology, statistics, and graph theory. Social network analysis is now one of the major paradigms in contemporary sociology, and is also employed in a number of other social and formal sciences. Together with other complex networks, it forms part of the nascent field of network science.

Social networking is discussed in [15]. Social networking refers to using internet-based social media sites to stay connected with friends, family, colleagues, or customers. Social networking can have a social purpose, a business purpose, or both through sites like Facebook, X, Instagram, and Pinterest. Social networking is also a significant opportunity for marketers seeking to engage customers. Marketers use social networking to increase brand recognition and encourage brand loyalty. Social media can help connect people with businesses for various needs. There are disadvantages related to social media, including spreading misinformation, concerns posed by user anonymity, and the high cost of using and maintaining social network profiles. Marketers commonly use social networking to increase brand recognition and encourage brand loyalty. Social media marketing helps promote a brand's voice and content.

Social network analytics is discussed in [16]. Social media websites can effectively recommend friends and connections to them. Some social media websites like linkedin also notify the jobs you may be interested to apply. For doing that, they study the social networks, consisting of a set of objects (nodes) with interconnections (edges), where people act as nodes of the network and determine how connected the two profiles are by various parameters. And this helps them in recommending the jobs, connections etc. Social network analysis can also help prevent spreading of rumors on social media by detecting the important nodes in the social network which when turned off will minimize the spread of a particular rumor. Social network analytics have great application in the data science industry and it can be used in a wide range of applications such as transportation network.

Most popular social networking sites are discussed in [17]. Top 22 social networking Sites include the following: LinkedIn, Facebook, Tumblr, Instagram, Reddit, YouTube, TikTok, Twitter, WhatsApp, Quora, Medium, Snapchat, Telegram, Goodreads, Pinterest, Discord, WeChat, Messenger. Twitch, Clubhouse, QQ, Kwai. With the launch of networking sites like Facebook, YouTube, and Twitter, the number of people actively engaging in social networking has risen. The statistics for 2023 show there are 4.48 billion active social media users worldwide. Marketers are incorporating new ways to increase their brand reach and get to the right audience. And what better way to find audience than to utilize the power of social networking? Social media strategy not only requires platforms that have the most users, but these platforms should also serve your business needs well.

Social and economic networks are discussed in [18]. Networks of relationships help determine the careers that people choose, the jobs they obtain, the products they buy, and how they vote. The many aspects of lives that are governed by social networks make it critical to understand how they impact behavior, and which network structures are likely to emerge in a society. Provides empirical background on networks and the regularities that they exhibit, discusses random graph-based models and strategic models of network formation. It helps to understand behavior

Fig. 2.3 Social network

in networked societies, with a detailed analysis of learning and diffusion in networks, decision making by individuals who are influenced by their social neighbors, game theory and markets on networks, and a host of related subjects. Also describes the varied statistical and modeling techniques used to analyze social networks.

See Fig. 2.3 symbolically related to social networks.

2.2.4 Battle and Military Networks

Battle Networks are discussed in [19]. Militaries use battle networks to detect what is happening on the battlefield, process that data into actionable information, decide on a course of action, communicate decisions among forces, act on those decisions, and assess the effectiveness of the actions taken. Battle networks are sometimes referred to as the "sensor-to-shooter kill chain", and they are widely acknowledged as an increasingly important element of modern warfare. While the importance of battle networks has garnered more attention in recent years, battle networks themselves are not new. Early battle networks used scouts, couriers, flags, telegraphs, and wired field telephones to transmit information and decisions among forces on the battlefield. More advanced battle networks began to emerge in World War II with the widespread adoption of technologies such as radar, sonar, radio communications, and aerial reconnaissance. As battle networks became faster, longer range, and more advantageous to militaries, the networks themselves also became an attractive target.

A 21st Century battle network [20] is a combination of target acquisition sensors, target localization sensors, command and control (C2) elements, weapons, weapon platforms, and the electronic communications linking them together. Prior to the advent of electronic communications and sensors, C2 and weapons had to be close enough to allow for rapid communication. As this was generally limited to the human voice, bugle calls, and, if weather and lighting conditions permitted, flag signals, dispersal options were constrained by the limits of human hearing and eyesight. Dispersing these functions in a network across a large area has a number of potential advantages and can greatly increase their effectiveness. Networking allows weapon platforms to engage targets they cannot otherwise detect or track by leveraging the capabilities of distant sensor systems. Networking functions can also improve overall effectiveness by giving commanders a broader view of the overall situation, as viewed from a number of disparate sensors.

The outlined new battle network is discussed in [21]. The Department of the Air Force has identified 50 programs that will make up the core of its contribution to the joint all-domain command and control concept, The DAF Battle Network programs, which span multiple offices and agencies across the department, continue the evolution of the Advanced Battle Management System, the Air Force's long-developing, ambitious attempt to replace a battle management aircraft with a network of systems and capabilities connecting sensors and shooters around the world. ABMS investment is slated to accelerate, projecting $3.7 billion in research and development spending through fiscal 2028. ABMS is therefore not just a weapon system platform or sensor, it is the aggregate of materiel and non-materiel solutions to integrate the essential data network that connects and empowers current and future weapon system platforms and sensors to fight and win in the modern era.

Challenges of military networks and technological developments are discussed in [22]. Many nations are currently at different stages of implementing digital modernization of their armed forces. Elbit Systems partners with MoDs to jointly realize the vision of a multi-dimensional networked combat C4ISR system. The heart and enabler of this C4ISR system is the tactical network—an integrated mobile and deployed IP WAN, interconnecting all combat and command, land, air and sea elements in the theatre, and connecting to the upper level fixed network and adjacent coalition WANs. The C4ISR Network needs to be capable in terms of capacity and latency; it must be efficient in its use of bandwidth and spectrum and adaptive to changes in network conditions. The network has to be secure against EW and cyber threats as well as manageable—able to be set up and used by staff with limited technical expertise. It should be interoperable with upper echelon, adjacent coalition and existing legacy networks, and standard in its interfaces to simplify integration.

The problems of building a safety net are discussed in [23]. The disarmament, demobilization, and reintegration of ex-combatants has become an integral part of peace building. Although the main purpose of such interventions is to dissolve the military structures of armed groups, there is growing evidence that ex-combatant networks often remain intact. The paper investigates why such structures continue to

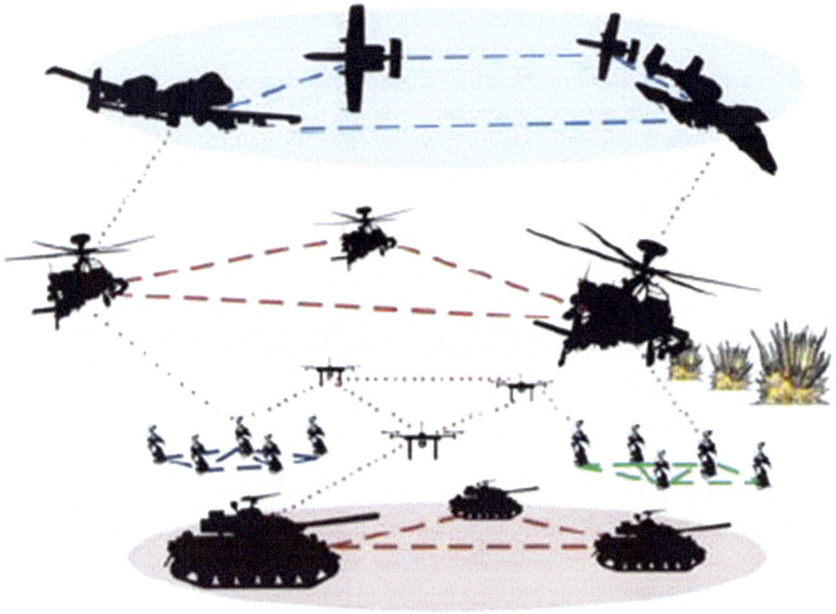

Fig. 2.4 Battle networks

thrive, and argues that ex-military networks are stronger when ex-commanders have weak links to elite patronage systems. Ex-combatants who are unable to rely on their former superiors for economic assistance must instead build denser ties to each other to gain access to a social safety net. To assess the argument, the paper conducts a comparative social network analysis (SNA) of two ex-military networks. This innovative approach helps to uncover previously overlooked, but central, dynamics related to ex-combatant groups. It is shown that SNA provides a range of underutilized tools and exact definitions that can increase our understanding of ex-military networks.

See Fig. 2.4 symbolically related to military and battle networks.

2.2.5 *Economic Networks*

Economic networks [24] are combinations of individuals, groups or countries that interact to benefit the entire community. The main goal of the group in an economic network is to strengthen its position in a market. Economic networks use all available competitive advantages and resources of each member to increase the output and wealth of the entire group. The composition of these networks may vary. In some economic networks members are immutable, whereas in other networks they can freely join or leave. Economic networks can take different forms and consist of groups of individuals, companies or countries sharing a common goal. Common types of economic networks can appear in the form of joint ventures between two or more companies, partnerships between corporations (especially in different countries). Activities in economic networks can include recruiting, surveying, learning, and sharing resources. Benefits also cover a larger labor supply, cost savings, and knowledge sharing.

An economic network [25] is a combination of individuals, groups, or countries interacting to benefit the community as a whole. The primary goal of the group in an economic network is to strengthen its position in a market. Common types of economic networks are joint ventures between two or more companies or partnerships between corporations. The advantages of an economic network are access to a larger labor pool of talent and cost savings. The disadvantage is that it may result in an imbalance of power between larger members and smaller ones. Economic networks use all available competitive advantages and resources of each member to increase the production and wealth of the entire group. The composition of these networks may vary. In some economic networks, membership may be static (where members do not change), while in others, the network may be dynamic. The activities in networks may consist of any number of things including recruitment, surveying, knowledge, and resource sharing.

Networks for innovative and resilient economies are discussed in [26]. Firm networks affect innovation and growth through diffusion of knowledge and information. Social and economic networks affect innovation and economic growth through facilitating diffusion of knowledge and information. However, networks are associated with externalities, as their initial costs are borne mostly by network organizers while their benefits can be shared with network members. Policy interventions may be necessary to develop networks to the optimal level and maximize social welfare. Which policy measures are effective in creating firm networks is not evident because it is not clear what types of networks contribute to knowledge diffusion and innovation. Strong and dense networks with trust within regional or organizational communities lead to more effective knowledge diffusion and innovation. But weak ties with outsiders can be more effective sources of information because the knowledge of outsiders is more diversified than of partners within the community.

Understanding economy by networks is in [27]. The purpose is to survey studies for estimating and analyzing different types of economic networks. The focus is on data-based approaches allowing direct estimation of networks from empirical data without the need of relying on theoretical assumptions. As there is a large variety of different economic networks, e.g., interbank, investment, director, ownership, financial, product or trade networks, a systematic categorization is presented by meaning of the nodes within these networks. These can correspond to banks, firms, investors, products, stocks etc. Practical methods are reviewed for graphically exploring such networks and discuss useful databases for obtaining the empirical data for the computational construction of economic networks. There is growing interest in studying economy by means of network science, with realization that it cannot be investigated by constituting components only, and analyzing interplay between all relevant parts is extremely important.

Cluster networks in economy are discussed in [28]. Recent economic transformations in the world economy are progressing in two divergent directions—international production fragmentation and industrial agglomeration. Based on extensive data analysis and using models of interdependencies between key economies, this book analyses innovation systems that cross national borders. It is shown that technological complexity is an important factor in the formation of highly specific production networks, and why, for a number of production systems, fragmentation and clustering are two sides of the same coin. By outlining the picture of a world economy structured around networks of clusters and joined together through systems of linkages of components, people and knowledge flows, the work helps to promote a better understanding of recent economic transformations.

Economy in the 21st century is discussed in [29]. The first two decades of the 21st century have been a time of technological and economic progress. A billion people have come out of poverty and half the world will be in the consuming class. In the advanced economies, aggregate economic data show economic and job growth, with employment at record highs in some countries. Yet the public mood in many of these economies is divergent, differing by socio-economic group and varying from place to place, ranging from satisfaction and optimism, to discontent about the present and pessimism about the economic future. Opportunities for work have expanded, employment rates have risen to record levels in many countries, and many benefits have improved, although not everywhere. At the same time, work polarization and income stagnation, while varying in magnitude across countries, have grown. While availability and cost of many discretionary goods and services have fallen sharply, the cost of basic necessities such as housing, healthcare, and education has grown.

See Fig. 2.5 symbolically related to economic networks.

2.2.6 Distribution Networks

A distribution network in a supply chain [30] is an interconnected group of storage facilities and transportation systems that receive inventories of goods and then deliver them to customers. It is an intermediate point to get products from the manufacturer to the end customer, either directly or through a retail network. A fast and reliable distribution network is essential in today's instant gratification society of consumers. In a supply chain, a distribution network is an interconnected group of storage facilities and transportation systems that receive inventories of goods and then deliver them to customers. It is an intermediate point to get products from the manufacturer to the end customer, either directly or through a retail network. There are many types of distribution networks, such as a hub-and-spoke or decentralized, that work best for different types of products. A fast and reliable distribution network is essential in today's instant gratification society of consumers.

A distribution network can be seen as the flow of goods [31] from a producer or supplier to an end consumer. The network consists of storage facilities, warehouses, and transportation systems that support the movement of goods until they reach

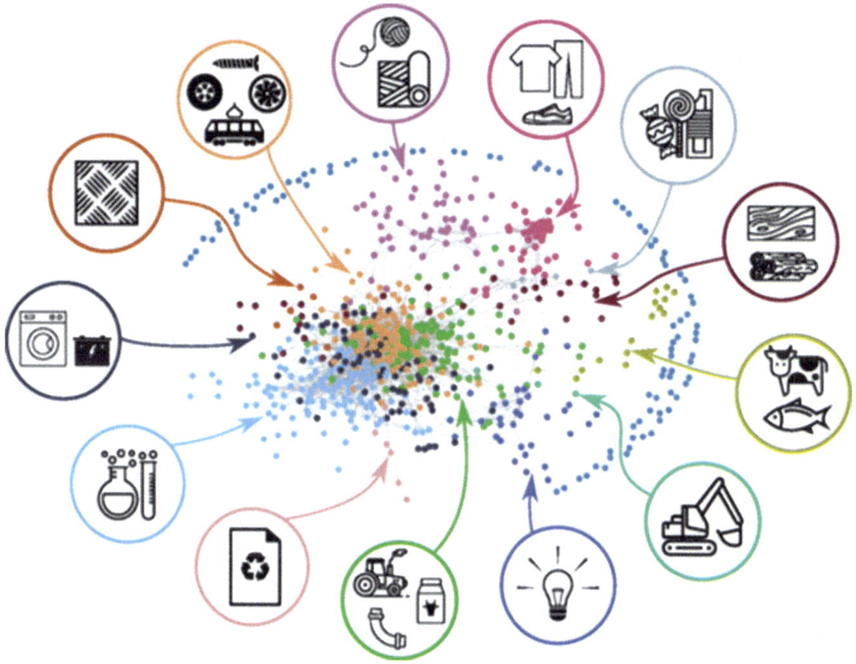

Fig. 2.5 Economic network

the end consumer. The process of ensuring the consumer receives the product from the manufacturer is done through direct sales or by following a retail network. A distribution network can be seen as the flow of goods from a producer or supplier to an end consumer. When defining the structure of a distribution network, the most crucial factors are the product demands of the end customer, customer experience, product variety and product availability, response time, and product returnability. Depending on the size of an enterprise or business, distribution networks vary in structure and size. Companies like Amazon or Apple are likely to own more sophisticated and complicated distribution networks, transportation, and logistics systems. Distribution networks transform over time as businesses expand and aim to reach more consumers.

Types of distribution networks in supply chain management (SCM) are discussed in [32]. The main goal of SCM is to improve the overall efficiency of the supply chain by reducing costs and time while improving quality and service. SCM must manage four key elements: *Integration*, where all activities must be coordinated and integrated both within each company as well as between different companies in order to avoid duplication of effort or conflicting objectives; *Sourcing*, where the right materials must be sourced from the right suppliers at the right time in order to meet demand; *Conversion,* with raw materials converted into finished products efficiently in order to meet customer demand; *Logistics*, where finished products must be delivered to customers on time and at a reasonable cost. Understanding the different types of distribution networks available can help businesses make informed decisions regarding their supply chain management strategies while meeting their customer demands efficiently at a lower cost point through effective procurement practices.

Benefits of a distribution network are discussed in [33] which include: *Cost reduction*, with gaining considerate savings in valuable resources, involving time, business assets, and personnel efforts; *Accuracy*, where advanced distribution management software systems will allow the organization or company to order the exact number of products that are much-needed; *Minimum use of resources*, where the distribution management system will lead to the optimum utilization of resources of the business organization or company, being the best thing about a powerful distribution network; *Simple, easy, and enhanced monitoring*, where many business organizations, industries and companies are able to monitor transactions and tasks of their business very easily; *Time management*, where with simple and creative monitoring and maintenance methods, many business organizations and industries will be able to save a lot of time, resources, and effort; *Relationship management*, in order to maintain effective relationship among manufacturers and distributors.

Optimal operation of active distribution networks [34] provides case studies, modern implementations and supporting flowcharts and code, along with current research in congestion management, service restoration and voltage control of active distribution networks. Contains an overview of the active distribution network concept, presents key approaches for the congestion management of active distribution networks, and cover approaches in uncertainty, scheduled re-profiling, and more. Reviews how to optimally operate active distribution networks under normal conditions and demonstrates worked solutions and contingency planning to mitigate unforeseen challenges. Provides clear guidance on optimizing congestion management, and demonstrates how to implement distributed voltage control in systems using active distribution networks. Also provides an extensive body of methods, associated case studies, worked solutions and implementation discussions on how to embed best practices in engineering and research workflows.

See Fig. 2.6 symbolically related to distribution networks.

Fig. 2.6 Distribution network

2.2.7 Virtual Networks

A virtual network [35] is a network where all devices, servers, virtual machines, and data centers that are connected are done so through software and wireless technology. This allows the reach of the network to be expanded as far as it needs to for peak efficiency, in addition to other benefits. A local area network, or LAN, is a wired network that can usually only reach within the domain of a single building. A wide area network, or WAN, is another kind of wired network, but the computers and devices connected to the network can stretch over a half-mile in some cases. Conversely, a virtual network doesn't follow conventional rules of networking because it isn't wired at all. All devices that interact with each other in the network do so through internet technology, allowing them to have a further reach than they would have if they were wired. The network itself is as limitless as the internet. There are three classes of virtual networks: VPN standing for virtual private network, VLAN using partitions to group devices on a LAN, and VXLAN as virtual extensible LAN.

Virtual networking [36] enables the communication between multiple computers, virtual machines, virtual servers, or other devices across different office and data center locations. While physical networking connects computers through cabling and other hardware, virtual networking extends these capabilities by using software management to connect computers and servers over the internet. It uses virtualized versions of traditional network tools, like switches and network adapters, allowing for more efficient routing and easier network configuration changes. Virtual networking enables devices across many locations to function with the same capabilities as a traditional physical network. This allows for data centers to stretch across different physical locations and gives network administrators new and more efficient options, like the ability to easily modify the network as needs change, greater flexibility in provisioning the network to specific needs and applications, and the capacity to move workloads across the network infrastructure.

Network virtualization is discussed in [37]. In computing, network virtualization is the process of combining hardware and software network resources and network functionality into a single, software-based administrative entity, a virtual network. Network virtualization involves platform virtualization, often combined with resource virtualization. Network virtualization is categorized as either external virtualization, combining many networks or parts of networks into a virtual unit, or internal virtualization, providing network-like functionality to software containers on a single network server. Various equipment and software vendors offer network virtualization by combining any of the following: Network hardware, such as switches and network adapters; Network elements, such as firewalls and load balancers; Networks, such as virtual LANs (VLANs); Network storage devices; Network machine-to-machine elements; Network mobile elements, such as laptop computers, tablet computers, and smartphones; Network media, such as Ethernet and Fibre Channel.

Pluralistic approach to virtual networks is discussed in [38] which concerns virtualization techniques that allow sharing computational resources, slicing a real

Fig. 2.7 Virtual networks

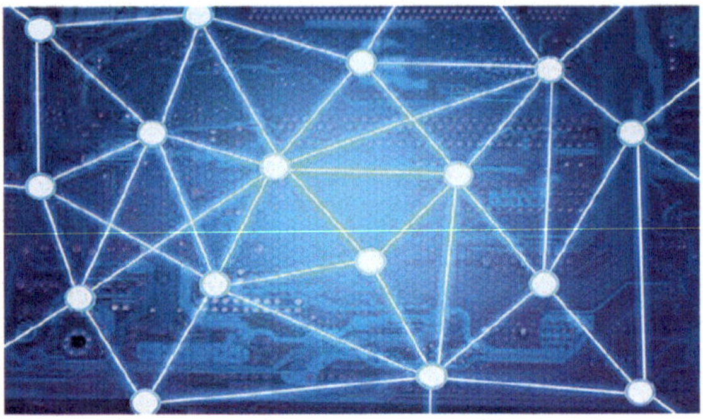

computational environment into virtual computational environments that are isolated from one another. Virtualization platforms performance analysis is provided. Defines the primitives that the network virtualization infrastructure must provide for allowing the piloting plane to manage virtual network elements. To control and manage virtual network elements, five primitives that the network virtualization infrastructure must provide are defined as: instantiate, delete, migrate, monitor and set. Then it moves to survey existing control algorithms for virtual networking, also describes the main challenges for packet forwarding using a virtualization tool and describes a proposal for local control of virtual networks. Within each physical node, this proposal guarantees the service level acquired by each virtual network, even in the presence of misbehaving virtual networks.

Fundamentals of virtual networking are discussed in [39], where virtual networking is a technology that facilitates the control of one or more remotely located computers or servers over the Internet. Here data communication occurs between two or more virtual machines. It is similar to traditional networking but interconnection occurs through virtual computing environment means in virtual networking all devices, servers, virtual machines are connected through software and wireless technology. It is based on physical networking principle but its functions are mostly software-driven. In virtual networking devices across many locations connect with each other and function with the same capabilities as a traditional physical network. Virtual networking reduces the cost and complexity of operating and maintaining hardware. Virtual networking software is installed on remote computers or servers to avail of the virtual networking facility.

See Fig. 2.7 symbolically related to virtual networks.

2.2.8 Neural Networks

A neural network is a method in artificial intelligence that teaches computers to process data in a way that is inspired by the human brain. It is a type of machine learning process, called deep learning that uses interconnected nodes or neurons in a layered structure that resembles the human brain.

A neural network [40] is a neural circuit of biological neurons, sometimes also called a biological neural network, or a network of artificial neurons or nodes in the case of an artificial neural network. Artificial neural networks are used for solving artificial intelligence (AI) problems; they model connections of biological neurons as weights between nodes. A positive weight reflects an excitatory connection, while negative values mean inhibitory connections. All inputs are modified by a weight and summed. This activity is referred to as a linear combination. Finally, an activation function controls the amplitude of the output. For example, an acceptable range of output is usually between 0 and 1, or it could be -1 and 1. These artificial networks may be used for predictive modeling, adaptive control and applications where they can be trained via a dataset. Self-learning resulting from experience can occur within networks, which can derive conclusions from a complex and seemingly unrelated set of information.

Neural network in machine learning is discussed in [41]. Neural network is the fusion of artificial intelligence and brain-inspired design that reshapes modern computing. With intricate layers of interconnected artificial neurons, these networks emulate the intricate workings of the human brain, enabling remarkable feats in machine learning. There are different types of neural networks, from feedforward to recurrent and convolutional, each tailored for specific tasks. Neural networks mimic the basic functioning of the human brain and are inspired by how the human brain interprets information. They solve various real-time tasks because of the ability to perform computations quickly. Artificial neural network has a huge number of interconnected processing elements, also known as nodes. These nodes are connected with other nodes using a connection link. The connection links have weights which contain the information about the input signals. Each iteration and input in turn leads to modification of these weights.

Neural network types and applications are discussed in [42]. Neural networks use algorithms to mimic the workings of the human brain to process and find relationships in datasets. They process data more efficiently and feature improved pattern recognition and problem-solving capabilities when compared to traditional computers. The architecture of a neural network comprises node layers that are distributed across an input layer, single or multiple hidden layers, and an output layer. Nodes are 'artificial neurons' linked to each other and are associated with a particular weight and threshold. Once the output of a single node crosses its specified threshold, that particular node is activated, and its data is transmitted to the next layer in the network. If the threshold value of the node is not crossed, data is not transferred to the next network layer. Known types of neural networks include: Convolutional neural networks, Recurrent neural networks, Feed-forward neural networks, Modular neural networks, and Generative adversarial networks.

Artificial neural network is discussed in [43]. An artificial neural network is a biologically inspired computational model that is patterned after the network of neurons present in the human brain. Artificial neural networks can also be thought of as learning algorithms that model the input–output relationship. Applications of artificial neural networks include pattern recognition and forecasting in fields such as medicine, business, pure sciences, data mining, telecommunications, and operations managements. An artificial neural network transforms input data by applying a nonlinear function to a weighted sum of the inputs. The transformation is known as a neural layer and the function is referred to as a neural unit. The intermediate outputs of one layer, called features, are used as the input into the next layer. The neural network through repeated transformations learns multiple layers of nonlinear features (like edges and shapes), which it then combines in a final layer to create a prediction (of more complex objects).

Graph neural networks are discussed in [44]. Deep learning models are at the core of artificial intelligence research today. It is well known that deep learning techniques are disruptive for Euclidean data, such as images or sequence data, and not immediately applicable to graph-structured data such as text. This gap has driven a wave of research for deep learning on graphs, including graph representation learning, graph generation, and graph classification. The new neural network architectures on graph-structured data (or GNN) have performed remarkably on these tasks, demonstrated by applications in social networks, bioinformatics, and medical informatics. GNN still face many challenges ranging from the foundational methodologies to the theoretical understandings of the power of the graph representation learning. It discusses the goals of graph representation learning and reviews the history, current developments, and future directions of GNN; presents and reviews fundamental methods and theories concerning GNN.

See Fig. 2.8 symbolically related to neural networks.

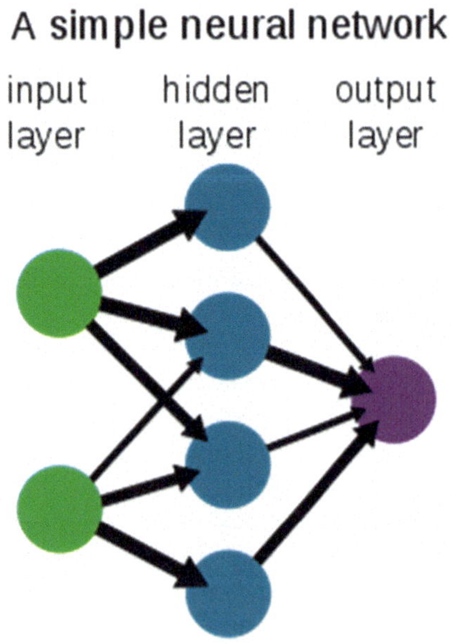

Fig. 2.8 Simplified view of a feed-forward artificial neural network

2.2.9 Psychological Networks

A brief summary on the rapidly growing network-related publications, especially in such areas as psychology and psychopathology, with their main ideas and contents is provided below.

Network analysis in psychological science is discussed in [45]. In recent years, network analysis has been applied to identify and analyze patterns of statistical association in multivariate psychological data. In these approaches, network nodes represent variables in a data set, and edges represent pairwise conditional associations between variables in the data, while conditioning on the remaining variables. The paper provides an anatomy of these techniques, describes the current state of the art and discusses open problems. Identifies relevant data structures in which network analysis may be applied: cross-sectional data, repeated measures and intensive longitudinal data; discusses the estimation of network structures in each of these cases, as well as assessment techniques to evaluate network robustness and replicability. Successful applications of the technique in different research areas are highlighted, and limitations and challenges for future research discussed. Network approaches involve the identification of system components and the relations among them.

Visualization of psychological networks is discussed in [46]. Networks have emerged as a popular method for studying mental disorders. Psychopathology networks consist of aspects (e.g., symptoms) of mental disorders (nodes) and the connections between those aspects (edges). Unfortunately, the visual presentation of networks can occasionally be misleading. For instance, researchers may be tempted to conclude that nodes that appear close together are highly related, and that nodes that are far apart are less related. This is not always the case. In networks plotted with force-directed algorithms, the most popular approach, the spatial arrangement of nodes is not easily interpretable. However, other plotting approaches can render node positioning interpretable. The paper provides a brief tutorial on several methods including multidimensional scaling, principal components plotting, and eigenmodel networks. It compares the strengths and weaknesses of each method, noting how to properly interpret each type of plotting approach.

Network analysis in psychology is discussed in [47]. The main goal of this work is to introduce a new approach called network analysis for its application in the field of psychology. This paper presents the network model in a brief and simple way and, as far as possible, away from technicalities and the statistical point of view. The aim is, on the one hand, to take the first steps in network analysis, and on the other, to show the theoretical and clinical implications underlying this model. Firstly, the roots of this approach are discussed as well as the way it has of understanding psychological phenomena, specifically psychopathological problems. The concepts of network, node and edge, the types of networks and the procedures for their estimation are all addressed. Next, measures of centrality are explained and some applications in the field of psychology are mentioned. Later, this approach is exemplified with a specific case, which estimates and analyzes a network of personality traits. The syntax of this analysis is provided.

Person-specific networks in psychopathology are discussed in [48]. In the psychological network approach, mental disorders such as major depressive disorder are conceptualized as networks. The network approach focuses on the symptom structure or the connections between symptoms instead of the severity (i.e., mean level) of a symptom. To infer a person-specific network for a patient, time-series data are needed. By far the most common model to statistically model the person-specific interactions between symptoms or momentary states has been the vector autoregressive (VAR) model. Although the VAR model helps to bring psychological network theory into clinical research and closer to clinical practice, several discrepancies arise when we map the psychological network theory onto the VAR-based network models. These challenges and possible solutions are discussed in this review.

Network analysis and its applications in psychology are discussed in [49]. Network analysis models have been widely used in psychology research in recent years. Unlike latent variable models which conceive observable variables as outcomes of unobservable latent factors, network analysis models apply the graph theory to construct a network to depict the associations among observable variables. The observable variables are treated as nodes and the associations between them are treated as edges. As such, network analysis models reveal the relationships among observable variables and the dynamic system resulted from the interactions between these observable variables. With indices reflecting individual nodes' characteristics (such as centrality) and network structural characteristics (such as small-worldness), network analysis models provide a new perspective for visualization and for studying various psychological phenomena. In the past decade, network analysis models have been applied in the fields of personality, social, and clinical psychology as well as psychiatry.

See Fig. 2.9 symbolically related to psychological networks.

2.2.10 Criminal Networks

Explanation of criminal networks is in [50] which offers some remarks on a few critical issues related to explanation in criminal network research. It first discusses two distinct perspectives on networks, namely a substantive approach that views networks as a distinct form of organization, and an instrumental one that interprets networks as a collection of nodes and attributes, the latter standing at the basis of Social Network Analysis. Social network techniques can be applied to investigate criminal networks while rejecting the notion of networks as a distinct form of organization. The article discusses some potential pitfalls associated with the instrumental approach and cautions against an over-reliance on structural measures alone when interpreting real-world networks. It stresses the need to complement these measures with additional qualitative evidence, and discusses the use of Quadratic Assignment Procedure regression models as a viable strategy to test hypotheses based on criminal network data.

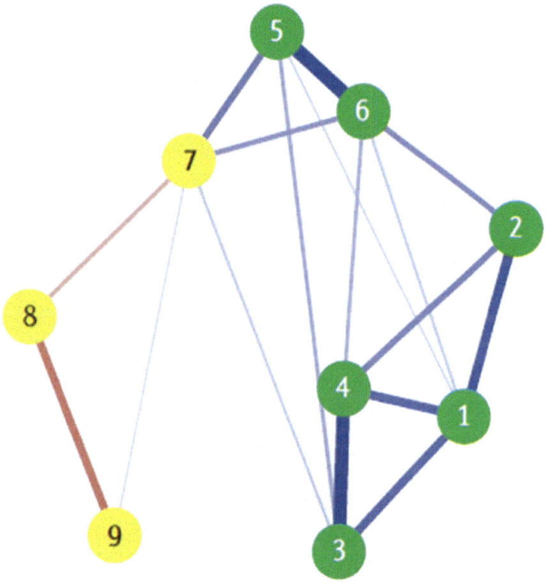

Fig. 2.9 Psychological network: stress symptoms 1–6, social symptoms 7–9

Understanding of criminal networks is discussed in [51]. The book is a short methodological primer for those interested in studying illicit, deviant, covert, or criminal networks using social network analysis. It is chock-full of graphics, checklists, software tips, step-by-step guidance, and straightforward advice. Covering all the essentials, it highlights three themes: the theoretical basis of networked criminology, methodological issues and useful analytic tools, and producing professional analysis. Also combines conceptual and empirical work with advice on designing networking studies, collecting data, and analysis. Relevant, practical, theoretical, and methodologically innovative, it promises understanding of how to cross over from conventional investigations of crime to the study of criminal networks.

Networks in criminology and justice studies are considered in [52]. This innovative collection of original essays showcases the use of social networks in the analysis and understanding of various forms of crime. More than any other past research endeavor, it applies to criminology the many conceptual and methodological options from social network analysis. Crime and Networks is the book that looks at the use of networks in understanding crime, and can be used for advanced undergraduate and beginner's graduate level courses in criminal justice and criminology.

New forms of organized crime using networks are discussed in [53]. A newer structure of organized crime occurs in the virtual world of cyberspace. Organized criminal groups use digital services but also run digital services, both for financial or other material benefit. Hackers threaten and extort victims, illicit drugs are sold and purchased online, as are falsified medical products, counterfeit merchandise, illicit wildlife specimen and stolen property. The Internet and the Darknet provide virtual spaces that can be used to connect criminals to customers without physical contacts.

Criminals associate when it is necessary to operate a criminal scheme. If cyber-related crimes permit criminals to become more disconnected from customers and victims, will members of organized criminal groups become more disconnected from each other? Continuing criminal associations may become less necessary. Instead, networks of convenience emerge in which criminals associate, perhaps temporarily, and only when it is necessary to do so.

Criminal networks and law enforcement is discussed in [54]. This collection presents an analysis of illicit networks and discusses implications for law enforcement and crime prevention. The contributors draw on a range of methodologies and apply them to diverse international criminological settings, from illegal fishing in the Indo-Pacific to 'money mule' networks in the Netherlands. Using a variety of examples, the book elucidates how and why criminals form networks of cooperation and how they can be disrupted. It is expected to be of interest to those who study criminology or criminal law, as well as law enforcement practitioners.

Detecting and disrupting criminal organizations is discussed in [55]. Transnational Crime Organizations try to obfuscate their structures and hierarchies. This creates direct challenges to designing analytical tools for recreating network structures from uncertain data, and tools for planning interdiction strategies. This research addresses these challenges by creating prototype software that quantifies the uncertainty of a network structure, and alternative interpretations of noisy data by assigning to each alternative network the probability of being correct. The prototyped analytic software enables analysts to create the most reliable structures of the network, compute this network uncertainty and its costs measured in uncertainty of the interdiction based on such a network. This patented technology can analyze partially-known criminal networks to reveal probable unknown criminals and determine the best strategy for interdiction. The new tools can help investigators to detect currently-unknown criminals and allocate resources to disrupt criminal networks.

See Fig. 2.10 symbolically related to criminal networks.

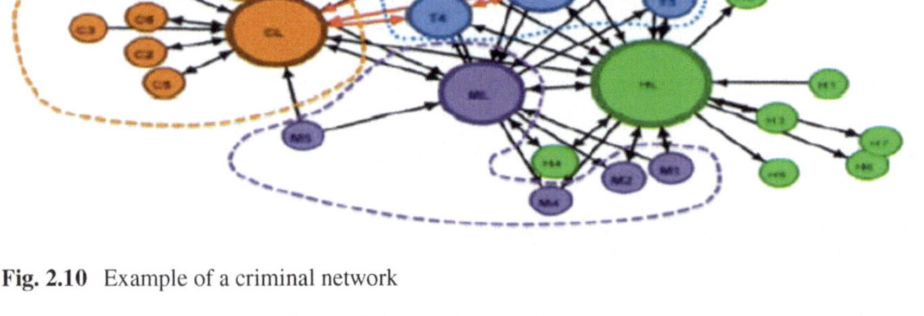

Fig. 2.10 Example of a criminal network

2.3 Network Analysis

An overview of network analysis is presented in [56], where a network is considered as a collection of interconnected entities or nodes that can communicate or exchange information with one another. Networks can be found in many different contexts, including social, technological, and biological systems. In a *social network*, the nodes might be individuals, and the connections between them might represent relationships such as friendship, family ties, or professional collaboration. In a *technological network*, the nodes might be devices such as computers or smartphones, and the connections between them might represent internet or cellular connections. In a *biological network*, the nodes might be cells or proteins, and the connections between them might represent chemical or physical interactions. Network analysis is a method of studying the relationships between entities in a network. It involves analyzing the connections, or links, between the entities, as well as the characteristics of the entities themselves. Network analysis can be used to study a wide range of systems, including social networks, transportation networks, and biological networks.

Structural network analysis in social sciences is discussed in [57]. The size and availability of network information has exploded over the last decade. Social scientists now share the stage of network analysis with computer scientists, physicists, and statisticians. While a number of introductions to network analysis are now available, most focus on theory, methods, or application alone. This book integrates all three. Network Analysis is an introduction to both the why and how of Social Network Analysis (SNA). It presents a broad theoretical overview rooted in social scientific approaches and guides users in how network analysis can answer core theoretical questions. It provides a comprehensive overview of descriptive and analytical approaches, including practical tutorials in R with sample data sets. Using an integrated approach, this book aims to quickly bring novice network researchers up to speed while avoiding common programming and analysis mistakes so that they might gain insight into the fundamental theories, key concepts, and methodological application of SNA.

Analysis of transport networks *is* overviewed in [58]. A transport network, or transportation network, is a network or graph in geographic space, describing an infrastructure that permits and constrains movement or flow. Examples include but are not limited to road networks, railways, air routes, pipelines, aqueducts, and power lines. The digital representation of these networks, and the methods for their analysis, is a core part of spatial analysis, geographic information systems, public utilities, and transport engineering. Network analysis is an application of the theories and algorithms of graph theory and is a form of proximity analysis. A wide range of methods, algorithms, and techniques have been developed for solving problems and tasks relating to network flow. Some of these are common to all types of transport networks, while others are specific to particular application domains.

Centrality in transportation network design is discussed in [59]. Efficient and effective transportation networks are backbones to modern societies. Methodologically, their design has mainly been driven by optimization approaches oftentimes with a strong cost focus. Their strategic planning, however, should go beyond detailed cost analysis and identify other key decision drivers. Transportation network centrality describes the appearance of a network; hence is crucial for network design. The book develops a strategic approach to transportation network design by conceptualizing transportation network centrality and relating it to the performance and quality of transportation networks. Consequently, the concept of network centrality serves to support decisions in strategic network design. A practical implementation of this approach is provided, demonstrating its feasibility.

Analyzing network structures and their stability is considered in [60]. A systematic, innovative introduction to the field of network analysis, provides a comprehensive overview of and guide to both the theoretical foundations of network psychometrics as well as modeling techniques developed from this perspective. It showcases cutting-edge methods in an easily accessible format, accompanied by problem sets and code. After working through this book it will be able to understand the theoretical foundations behind network modeling, infer network topology, and estimate network parameters from different sources of data. This book features an introduction on the statistical programming language R that guides on how to analyze network structures and their stability. While written in the context of social and behavioral science, the methods introduced are widely applicable to data sets from related fields of study. Additionally, while the text is written in a non-technical manner, technical content is highlighted in special textboxes.

Queries and storage methods for network databases are analyzed in [61], which provides a collection of concepts, algorithms, and techniques that effectively harness the power of Spatial Network Big Data. This book provides a first step towards understanding the immense challenges and novel applications of SNBD database systems. It explores these challenges via investigating scalable graph-based query processing strategies and I/O efficient storage and access methods. This book is of benefit to academics, researchers, engineers with a particular interest in network database models, network query processing, and physical storage models.

Brain network dysfunction in neuropsychiatric illness is discussed in [62]. Brain network function and dysfunction is the dominant model for understanding how the brain gives rise to normal and abnormal behavior. Moreover, neuropsychiatric illnesses continue to resist attempts to reveal an understanding of their bases. This volume provides a synthesis of the uses of multiple analytic methods as they are applied to neuroimaging data, to seek understanding of the neurobiological bases of psychiatric illnesses, understanding that can subsequently aid in their management and treatment. A principle focus is on the analyses and application of methods to functional magnetic resonance imaging data, which remains the most widely used neuroimaging technique for estimating brain network function, and several of the methods covered can estimate brain network dysfunction in resting and task-active states.

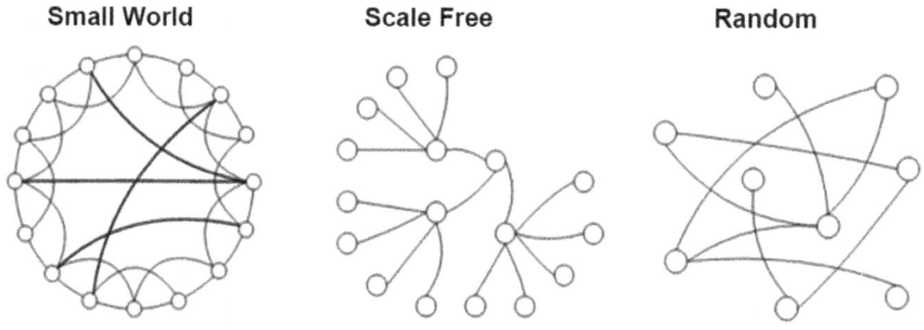

Fig. 2.11 Some types of real-world networks

Social network analysis [63] is the process of investigating social structures through the use of networks and graph theory. The paper leads to the theory of social networks, with a short introduction to graph theory and information spread. Real-world networks have a unique structure which often differs from random mathematical networks, with some features as follows. *Small World* phenomenon claims that real networks often have very short paths (in terms of number of hops) between any connected network members. This applies for real and virtual social networks and for physical networks such as airports or electricity of web-traffic routings. *Scale Free networks* with power-law degree distribution have a skewed population with a few highly-connected nodes and a lot of loosely-connected nodes. *Homophily* is the tendency of individuals to associate and bond with similar others, which results in similar properties among neighbors.

See Fig. 2.11 symbolically related to network analysis.

2.4 Network Management

Definition and practices of network management are discussed in [64]. Network management refers to the process of administering and monitoring the network to keep things running smoothly. Network management systems (NMS) collect data from network devices to determine how they operate and function. It's difficult to do this without poking into various key aspects that work together to ensure an efficient IT infrastructure. These components include: *Network Administration* which is responsible for the daily management of a network; *Network Provisioning* which involves the allocation and configuration of network resources; *Network Security* which focuses on protecting the network from threats, including unauthorized access; *Network Automation* which streamlines network management tasks by using software tools; *Network Maintenance* where networks must be maintained regularly to keep them working; *Network Protocol* as the rules and standards that govern communication between devices on a network; *Network Analytics* which involves analyzing data from network devices to optimize network performance.

Network management, as discussed in [65], is the sum total of applications, tools and processes used to provision, operate, maintain, administer and secure network infrastructure. The overarching role of network management is ensuring network resources are made available to users efficiently, effectively and quickly. It leverages fault analysis and performance management to optimize network health. Why do we need network management? A network brings together dozens, hundreds or thousands of interacting components. These components will sometimes malfunction, may happen to be misconfigured, get over utilized or just fail. Enterprise network management software must respond to these challenges by employing the best suited tools required to manage, monitor and control the network. The importance of network management includes the following: it *minimizes costly network disruptions, improves productivity, improves network security*, and *maintains holistic view* of network performance.

Concepts of network management are also considered in [66], which refers to two related concepts. First is the process of configuring, monitoring, and managing the performance of a network. Second is the platform that IT and NetOps teams use to complete these ongoing tasks. Network management systems have evolved to help IT teams operate in more agile ways, incorporating advanced analytics, machine learning, and intelligent automation to continually optimize network performance. As organizations adapt to a more distributed workforce, these network management systems are increasingly deployed in cloud and hosted environments. Network management systems collect data from connected network devices such as switches, routers, access points, and client devices. They also give network administrators fine-grained control over how those devices operate and interact with one another. The data captured from these devices is used to proactively identify performance issues, monitor security and segmentation, and accelerate troubleshooting.

Network management, as in [67], is the process of orchestrating network traffic and data flow across the enterprise ecosystem using network monitoring, network security, network automation, and other tools. The primary purpose of network management is to deliver a secure, reliable, and high-performing network to end-users, including business users in the enterprise and end customers. International Standards Organization considers the following types of network management: *Network fault management*, to anticipate, detect, and resolve network faults to minimize downtime; *Network configuration management* expected to keep with data and traffic demands in a large enterprise; *Network accounting and utilization management*, to consider network resources and add to enterprise costs; *Network performance management* involving various tasks to boost service availability and bandwidth speeds; *Network security management*, which is vital for resilience, risk management, and success.

Introduction to enterprise network and security management is provided in [68] which describes a balanced understanding of introductory and advanced subjects in both computer networking and cybersecurity. Although much of the focus is on technical concepts, managerial issues related to enterprise network and security planning and design are explained from a practitioner's perspective. Because of the critical importance of cybersecurity in today's enterprise networks, security-related issues are explained throughout the book, and four chapters are dedicated to fundamental knowledge. Challenging concepts are explained so readers can follow through with careful reading. The book is oriented on self-studying or studying information systems or computer science in a classroom setting.

Asset management for road networks is considered in [69]. Road networks include an extensive set of physical assets, including pavement, bridges, tunnels, structures, geotechnical assets, traffic signals, intelligent transportation system devices, facilities, signs, and other traffic and safety features. Many road network organizations are looking for innovative new ideas and approaches to replace aging systems, simplify complex business processes and support a comprehensive data-driven approach to asset management. They are looking for solutions that are easy to use; seamlessly integrate with their financial, work management, geospatial and life-cycle analysis systems; provide comprehensive mobile capabilities; are based on modern technology platforms; address visualization and reporting needs. The paper provides a business process and solution framework within which to significantly improve the information collection, maintenance, and management of road network assets, and accelerate decision making processes on asset management programs.

Managing Airline Networks [70], considering design, integration and innovative technologies, is a fully comprehensive description of state-of-the-art network management practices at airlines. Designed as a compendium on current practices and future trends in the field, the book offers an instructive guide through the complex world of non-linear production systems. It discusses the impact of network management on airline resource planning and performance, and examines the interplay between network management and adjacent functions. The book includes a practical case study and is enriched with academic perspectives. Discussing upcoming trends in the sector, the book provides an outlook on advanced technologies that may play a role in next-generation network management.

A guide and practices of network management are considered in [71]. Most articles and resources are written with technical terms by or for an audience of experts. To complicate it, there are different specialties represented in network management. Network management includes all the hardware and software that are connected to the business. Network operations include *monitoring the network traffic*, *bandwidth*, *troubleshooting outages*, and *maintaining network security*. Any network issues that impact the end user need to be addressed as well. The following components are usually included in network management: *Network devices*, which include the actual devices, such as desktops and mobile devices; *Types of network*, being simply the type of connection you use in order to allow all of those devices to communicate effectively and safely; and *Topologies*, where topology in network management is the way your network is structured, acting as a sort of map to your network and should list all the devices associated with your network.

See Fig. 2.12 symbolically related to network management.

Fig. 2.12 Areas of network management

2.5 Basics of Spatial Grasp Model and Technology

We have chosen this Spatial Grasp paradigm to investigate its applicability as a possible universal approach and technique that could simultaneously cover very different networking areas, including those mentioned in the previous sections and provide important united solutions. Only main technology ideas are mentioned below, more can be found in the related patent [72], published books [73–81], latest journal papers [82–89], also previous publications on this paradigm (mentioning only few and of the earliest, due to their number and period covered) [90–145].

Within Spatial Grasp Model and Technology (SGT), a high-level operational scenario expressed in recursive Spatial Grasp Language (SGL), starting in any world point *propagates, covers, and matches the distributed environment in parallel wavelike mode*. as symbolically shown in Fig. 2.13, Such propagation can result in returning and analyzing the reached states and data, which may be arbitrarily remote, or used for launching more waves.

The SGL allows for organizing *direct space presence and operations* with unlimited powers and parallelism. Its universal recursive organization, with operational scenarios called *grasp*, can be expressed by a single formula:

$$\textbf{\textit{grasp}} \rightarrow \textbf{\textit{constant}}\ |\textbf{\textit{variable}}|\ \textbf{\textit{rule}}\ (\{\textbf{\textit{grasp}},\})$$

The *rule* expresses certain action, control, description or context accompanied with operands, which can be any *grasps* too. The rules, starting in certain points, can organize navigation of the world sequentially, in parallel, or any combinations

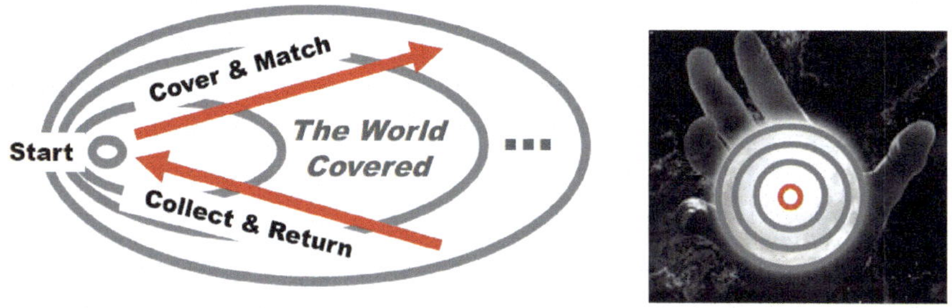

Fig. 2.13 Parallel wavelike world coverage under Spatial Grasp Model

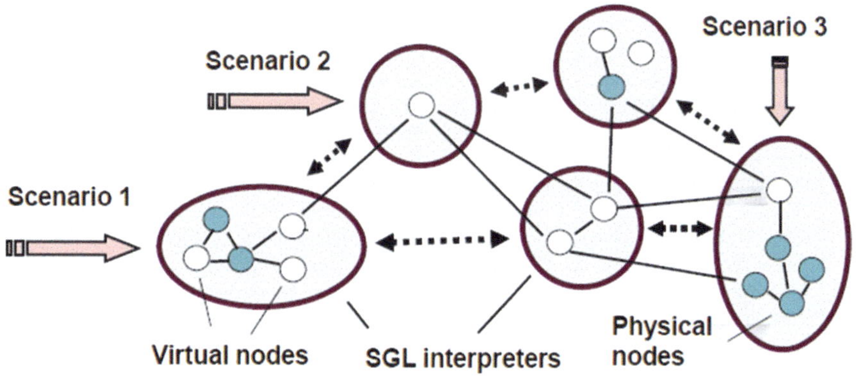

Fig. 2.14 SGL distributed networked interpretation

thereof. They can result in the same application points or cause movement to other world points with obtained results left there or returned. The *SGL interpreter* consists of specialized modules serving SGL scenarios or their parts happened to be inside this interpreter, also organizing exchanges with other interpreters for distributed SGL scenarios. Communicating SGL interpreters can be in arbitrary number of copies, representing *powerful spatial engines operating without central resources or control*, as in Fig. 2.14. They can effectively work with *spatial graph and network data* of any volumes and distributions. Details of SGT are explained in detail in Chaps. 3–5.

2.6 Conclusions

After review and analysis of networking approaches in very different areas of human activity, we feel the reason to investigate networking in most general form, as a universal method and tool for dealing with any kinds of systems, especially distributed. And even move higher, and try to find a superior philosophy, model, and technology that could describe, create, and manage any networking models, which could, in their turn, effectively cover any existing theoretical and practical systems. These ideas have lead us to the developed Spatial Grasp paradigm which already succeeded in numerous applications in very different areas and which is based on extremely flexible, independent, and recursive (flooding, smog, virus, or even soul-like) coverage of large distributed physical and virtual spaces. Also capable of creating in them any graph and network structures which may themselves become active and self-organized. In this book, we will be explaining and examining this idea and concept in detail, together with its practical applications.

References

1. Transport networks.: Inspire Registry (2015). https://inspire.ec.europa.eu/theme/tn
2. Transportation networks.: Travel forecasting resource. https://tfresource.org/topics/Transportation_networks.html
3. Durkin, J., Goodman, J., et al.: Building Multiservice Transport Networks (Networking Technology), 1st edn. Cisco Press (2012)
4. Murakami, M., Li, H., Ryoo, J.-D.: Packet transport networks: overview and future direction. In: APT/ITU Conformance and Interoperability Workshop (C&I-2), 2 edn. (2014). https://www.itu.int/en/ITU-T/C-I/interop/Documents/20140826/CI-2_INP-09_Packet_Transport_Networks_Overview_and_Future_Direction.pdf
5. Air Transport Network. https://en.wikipedia.org/wiki/Air_Transport_Network
6. Qureshi, K.N., Ul Islam, F., et al.: Improved road segment-based geographical routing protocol for vehicular ad-hoc networks. Electronics **9**(8), 1248 (2020). https://doi.org/10.3390/electronics9081248, https://www.mdpi.com/2079-9292/9/8/1248
7. Sujan.: What is Communication Network? Definition, Importance, and 4 Types (Explained) (2023). https://tyonote.com/communication_network/
8. Fitzek, F.H.P., Granelli, F., et al.: Computing in Communication Networks: from Theory to Practice, 1st edn. Academic Press (2020)
9. Srikant, R., Ying, L.: Communication Networks: an Optimization, Control, and Stochastic Networks Perspective, 1st edn. Cambridge University Press (2013)
10. Raghunandan, K.: Introduction to Wireless Communications and Networks: a Practical Perspective (Textbooks in Telecommunication Engineering), 1st edn. Springer (2023)
11. Lou, Y., Ahmed, N.: Underwater Communications and Networks (Textbooks in Telecommunication Engineering). Springer (2021)
12. Kaufman, C., Perlman, R., et al.: Network Security: private Communication in a Public World (Prentice Hall Series in Computer Networking and Distributed Systems), 3rd edn. Addison-Wesley Professional (2022)
13. Wu, Y., Singh, S., et al.: 6G Mobile Wireless Networks (Computer Communications and Networks), 1st edn. Springer (2022)
14. Social network. https://en.wikipedia.org/wiki/Social_network
15. Kenton, W.: What is Social Networking? Investopedia (2023). https://www.investopedia.com/terms/s/social-networking.asp#:~:text=Social%20networking%20refers%20to%20using,)%2C%20Instagram%2C%20and%20Pinterest.
16. Nanawati, S.: Social Network Analytics, Published in Analytics Vidhya (2019). https://medium.com/analytics-vidhya/social-network-analytics-f082f4e21b16
17. Singh, C.: 20+ Most Popular Social Networking Sites in 2024 (2023). https://www.socialpilot.co/social-networking-sites
18. Jackson, M.O.: Social and Economic Networks, Illustrated edn. Princeton University Press (2010)

19. Harrison, T.: Battle Networks and the Future Force. Center for Strategic and International Studies (2021). https://www.csis.org/analysis/battle-networks-and-future-force
20. Stillion, J., Clark, B.: What it takes to win: succeeding in 21st century battle network competitions. Center for Strategic and Budgetary Assessments (CSBA) (2015). https://csbaonline.org/uploads/documents/What-it-Takes-to-Win.pdf
21. Hadley, G.: DAF outlines a new 'Battle Network' as its contribution to JADC2. Air & Space Forces Magazine (2023). https://www.airandspaceforces.com/daf-battle-network-contribution-jadc2/
22. The Challenges of Military Networks and Technological Developments, Elbit Systems. https://elbitsystems.com/e-lynx_multi_channel/
23. Themnér, A., Karlén, N.: Building a safety net: explaining the strength of ex-military networks. Secur. Stud. **29**(2), 268–300. https://doi.org/10.1080/09636412.2020.1722851, https://www.tandfonline.com/doi/full/10.1080/09636412.2020.1722851
24. What is an Economic Network? Advantages and disadvantages of economic networks, Vietnambiz (2019). https://vietnambiz.vn/mang-luoi-kinh-te-economic-network-la-gi-uu-nhuoc-diem-cua-mang-luoi-kinh-te-20191011113428212.htm
25. Halton, C.: Economic network: what it is, pros and cons, examples (2022). https://www.investopedia.com/terms/e/economic-network.asp#:~:text=Key%20Takeaways-,An%20economic%20network%20is%20a%20combination%20of%20individuals%2C%20groups%2C%20or,companies%20or%20partnerships%20between%20corporations.
26. Todo, Y., Matous, P.: Economic networks for more innovative and resilient economies (2015). https://cepr.org/voxeu/columns/economic-networks-more-innovative-and-resilient-economies
27. Emmert-Streib, F., Tripathi, S., et al.: Understanding the world economy in terms of networks: a survey of data-based network science approaches on economic networks. Front. Appl. Math. Stat. Math. Finance **4** (2018). https://doi.org/10.3389/fams.2018.00037. https://www.frontiersin.org/articles/10.3389/fams.2018.00037/full
28. Wixted, B.: Innovation System Frontiers: cluster Networks and Global Value (Advances in Spatial Science). Springer (2009)
29. Manyika, J.: What the economy looks like so far for individuals in the 21st century (2020). https://www.linkedin.com/pulse/what-economy-looks-like-so-far-individuals-21st-century-james-manyika
30. Hayes, A.: Distribution network: definition, how it works, and examples, Investopedia (2021). https://www.investopedia.com/terms/d/distribution-network.asp#:~:text=In%20a%20supply%20chain%2C%20a,or%20through%20a%20retail%20network
31. Distribution Network.: The flow of goods from a producer or supplier to an end consumer. Written by CFI Team. https://corporatefinanceinstitute.com/resources/valuation/distribution-network/
32. What Are Types of Distribution Networks in Supply Chain Management? Oboloo FAQ's. https://oboloo.com/blog/what-are-types-of-distribution-networks-in-supply-chain-management/
33. Distribution Network—What is it and What are the Benefits? https://www.salesbabu.com/blog/what-is-a-distribution-network-and-its-benefits/
34. Wu, Q., Shen, F., et al.: Optimal Operation of Active Distribution Networks: congestion Management, Voltage Control and Service Restoration. Academic Press (2023)
35. Watts, S.: What is a virtual network? BMC (2020). https://www.bmc.com/blogs/virtual-network/
36. What is Virtual Networking? VMware by Broadcom. https://www.vmware.com/topics/glossary/content/virtual-networking.html
37. Network virtualization. https://en.wikipedia.org/wiki/Network_virtualization
38. Carlos, O., Duarte, M.B., Pujolle, G.: Virtual Networks: pluralistic Approach for the Next Generation of Internet (Networks and Telecommunications), 1st edn. Wiley-ISTE (2013).
39. Fundamentals of Virtual Networking. https://www.geeksforgeeks.org/fundamentals-of-virtual-networking/

40. Neural network. https://en.wikipedia.org/wiki/Neural_network
41. Anand, V.: Introduction to Neural Network in Machine Learning (2023). https://www.analyticsvidhya.com/blog/2022/01/introduction-to-neural-networks/
42. Ashtari, H.: What is a Neural Network? Definition, Working, Types, and Applications in 2022 (2022). https://www.spiceworks.com/tech/artificial-intelligence/articles/what-is-a-neural-network/
43. Artificial Neural Network.: NVIDIA Developer. https://developer.nvidia.com/discover/artificial-neural-network
44. Wu, L., Cui, P., et al.: Graph Neural Networks: Foundations, Frontiers, and Applications, 1st edn. Springer (2022)
45. Borsboom, D., Deserno, M.K., Rhemtulla, M., et al.: Network analysis of multivariate data in psychological science. Nat. Rev. Methods Primers **1**, 58 (2021). https://doi.org/10.1038/s43586-021-00055-w, https://www.nature.com/articles/s43586-021-00055-w
46. Jones, P.J., Mair, P., McNally, R.J.: Visualizing psychological networks: a tutorial in R. Front. Psychol. Sec. Quant. Psychol. Meas. **9** (2018). | https://doi.org/10.3389/fpsyg.2018.01742, https://www.frontiersin.org/articles/10.3389/fams.2018.00037/full
47. Fonseca-Pedrero, E.: Network analysis in psychology. Psychol. Pap. **39**(1), 1–12. https://doi.org/10.23923/pap.psicol2018.2852, https://www.papelesdelpsicologo.es/pdf/2852.pdf
48. Bringmann, L.F.: Person-specific networks in psychopathology: past, present, and future. Curr. Opin. Psychol. **41**, 59–64 (2021). https://www.sciencedirect.com/science/article/pii/S2352250X21000300?via%3Dihub
49. Yuqing, C., Shuyang, D., et al.: Network analysis and its applications in psychology. Adv. Psychol. Sci. **28**(1), 178–190 (2020). https://doi.org/10.3724/SP.J.1042.2020.00178, https://journal.psych.ac.cn/adps/EN/https://doi.org/10.3724/SP.J.1042.2020.00178
50. Campana, P.: Explaining criminal networks: strategies and potential pitfalls. Methodol. Innov. **9**, 1–10 (2016). https://doi.org/10.1177/2059799115622748. https://journals.sagepub.com/doi/pdf/https://doi.org/10.1177/2059799115622748
51. Bichler, G.: Understanding Criminal Networks: a Research Guide, 1st edn. University of California Press (2019)
52. Morselli, C.: Crime and Networks (Criminology and Justice Studies). Routledge (2013)
53. New forms of organized crime: networked structure.: UNODS, Unite Nations Office on Drugs and Crime. https://www.unodc.org/e4j/en/organized-crime/module-7/key-issues/networked-structure.html
54. Hufnagel, S., Moiseienko, A.: Criminal Networks and Law Enforcement: global Perspectives On Illegal Enterprise (Transnational Criminal Justice), 1st edn. Routledge (2019)
55. Detecting and Disrupting Transnational Criminal Organizations: analytics for Interdependent Smuggling and Money-Laundering Networks, Criminal Investigation and Network Analysis. DHS Centers of Excellence. https://cina.gmu.edu/projects/detecting-and-disrupting-transnational-criminal-organizations-analytics-for-interdependent-smuggling-and-money-laundering-networks/
56. What is Network Analysis—An overview, by Great Learning.: (2023). https://www.mygreatlearning.com/blog/what-is-network-analysis/#:~:text=Network%20analysis%20is%20a%20method,network%20functions%20as%20a%20whole
57. Rawlings, C.M.: Network Analysis (Structural Analysis in the Social Sciences). Cambridge University Press (2023)
58. Transport network analysis. https://en.wikipedia.org/wiki/Transport_network_analysis
59. Lange, A.: Centrality in Strategic Transportation Network Design: an Application to Less-than-Truckload Networks (Edition KWV). Springer Gabler (2019)
60. Isvoranu, A.-M., Epskamp, S., et al.: Network Psychometrics with R: a Guide for Behavioral and Social Scientists. Routledge (2022)
61. Yang, K.S., Shekhar, S.: Spatial Network Big Databases: queries and Storage Methods. Springer (2017)
62. Diwadkar, V.A., Eickhoff, S.B.: Brain Network Dysfunction in Neuropsychiatric Illness: methods, Applications, and Implications. Springer (2021)

63. Goldenberg, D.: Social Network Analysis: from Graph Theory to Applications with Python, Published in Towards Data Science (2021). https://towardsdatascience.com/social-network-analysis-from-theory-to-applications-with-python-d12e9a34c2c7
64. Iwuozor, J., Bottorff, C., Hoeper, M.: What is Network Management? Definition and Best Practices (2023). https://www.forbes.com/advisor/business/software/what-is-network-management/
65. What is Network Management?. https://www.opentext.com/what-is/network-management#:~:text=Network%20management%20is%20the%20sum,users%20efficiently%2C%20effectively%20and%20quickly.
66. What is Network Management? CISCO. https://www.cisco.com/c/en/us/solutions/enterprise-networks/what-is-network-management.html
67. BasuMallick, C.: What is network management? Definition, key components, and best practices, Spiceworks (2022). https://www.spiceworks.com/tech/networking/articles/what-is-network-management/
68. Shin, B.: A Practical Introduction to Enterprise Network and Security Management, 2nd edn. Auerbach Publications (2021)
69. Wilson, J., Trembath, M.: Asset Management for Road Networks, SAP (2021). https://rizing.com/wp-content/uploads/2021/08/SAP-Asset-Management-for-road-networks.pdf
70. Franke, M.: Managing Airline Networks: design, Integration and Innovative Technologies (Managing Aviation Operations). Routledge (2020)
71. Buhrkuhl, C.: Network Management: a Guide and Best Practices (2021). https://www.office1.com/blog/network-management-a-guide-and-best-practices
72. Sapaty, P.S.: A distributed processing system, European Patent N 0389655, Publ. 10.11.93, European Patent Office
73. Sapaty, P.S.: Mobile Processing in Distributed and Open Environments. Wiley, New York (1999)
74. Sapaty, P.S.: Ruling Distributed Dynamic Worlds. Wiley, New York (2005)
75. Sapaty, P.S.: Managing Distributed Dynamic Systems with Spatial Grasp Technology. Springer (2017)
76. Sapaty, P.S.: Holistic Analysis and Management of Distributed Social Systems. Springer (2018)
77. Sapaty, P.S.: Complexity in International Security: a Holistic Spatial Approach. Emerald Publishing (2019)
78. Sapaty, P.S.: Symbiosis of Real and Simulated Worlds under Spatial Grasp Technology. Springer (2021)
79. Sapaty, P.S.: Spatial Grasp as a Model for Space-Based Control and Management Systems. CRC Press (2022)
80. Sapaty, P.S.: The Spatial Grasp Model: applications and Investigations of Distributed Dynamic Worlds. Emerald Publishing (2023)
81. Sapaty, P.S.: Providing Integrity, Awareness, and Consciousness in Distributed Dynamic Systems. CRC Press (2024)
82. Sapaty, P.S.: Spatial management of air and missile defence operations. Math. Mach. Syst. (1) (2023). http://www.immsp.kiev.ua/publications/articles/2023/2023_1/01_23_Sapaty.pdf
83. Sapaty, P.S.: Providing distributed system integrity under spatial grasp technology. Math. Mach. Syst. **2** (2023). http://www.immsp.kiev.ua/publications/articles/2023/2023_2/02_23_Sapaty.pdf
84. Sapaty, P.S.: Providing global awareness in distributed dynamic systems. Int. Relat. Dipl. **11**(2), 87–100 (2023). https://doi.org/10.17265/2328-2134/2023.02.002. https://www.davidpublisher.com/Public/uploads/Contribute/6486c3d05a6cc.pdf
85. Sapaty, P.S.: Simulating distributed consciousness with Spatial Grasp Model. Math. Mach. Syst. (3) (2023)
86. Sapaty, P.S.: Network centrality operations under spatial grasp technology. J. Adv. Artif. Intell. Mach. Learn. **1**(1), 1–11 (2023). https://www.scivisionpub.com/journals/articleinpress-journal-of-advances-in-artificial-intelligence-and-machine-learning

87. Sapaty, P.S.: Managing distributed systems with spatial grasp patterns. Math. Mach. Syst. (4) (2023). http://www.immsp.kiev.ua/publications/articles/2023/2023_4/04_23_Sapaty.pdf
88. Sapaty, P.S.: Spatial networks as models for organoid cultures and brain research. Math. Mach. Syst. (1) (2024)
89. Sapaty, P.S.: Basic network operations under Spatial Grasp Model. Sci. Set. J. Phys. **3**(1), 01–16 (2024)
90. Bondarenko, A.T., Mikhalevich, S.B., Nikitin, A.I., Sapaty, P.S.: Software of BESM-6 computer for communication with peripheral computers via telephone channels. In: Computer Software, vol. 5. Institute of Cybernetics Press, Kiev (1970)
91. Sapaty, P.S.: A Method of organization of an intercomputer dialogue in the radial computer systems, in The Design of Software and Hardware for Automatic Control Systems. Institute of Cybernetics Press, Kiev (1973)
92. Sapaty, S.: A wave language for parallel processing of semantic networks. Comput. Artif. Intell. **5**(4), 289–314 (1986)
93. Sapaty, P.S.: Distributed modeling of cooperative behavior by mobile agents. In: Proceedings of the Sixth Conference on Computer Generated Forces and Behavioral Representation, pp. 599–613. IST UCF, Orlando, FL (1996)
94. Sapaty, P.S.: Distributed artificial brain for collectively behaving mobile robots. In: Proceedings of Symposium and Exhibition Unmanned Systems 2001, Baltimore, MD (2001)
95. Sapaty, P.S.: Crisis management with distributed processing technology. Int. Trans. Syst. Sci. Appl. **1**(1), 81–92 (2006). ISSN 1751-1461
96. Sapaty, P.S.: Global management of distributed EW-related system. In: Proceedings of International Conference Electronic Warfare: operations & Systems 2007, Thistle (2007)
97. Sapaty, P.S.: Distributed technology for global dominance. In: Proceedings of International Conference Defense Transformation and Net-Centric Systems 2008, as part of the SPIE Defense and Security Symposium, World Center Marriott Resort and Convention Center, Orlando, FL, USA (2008); Proceedings of SPIE—Volume 6981, Defense Transformation and Net-Centric Systems 2008, Raja Suresh, Editor, 69810T (2008)
98. Sapaty, P.S.: Distributed technology for global dominance. Keynote lecture. In: Proceedings of the Fifth International Conference in Control, Automation and Robotics ICINCO 2008, The Conference Proceedings, Funchal, Madeira, Portugal (2008)
99. Sapaty, P.S.: Human-Robotic Teaming: a Compromised Solution. AUVSI's Unmanned Systems North America 2008, San Diego, USA (2008)
100. Sapaty, P.S.: High-level communication protocol for dynamically networked battlefields. In: Proceedings of International Conference Tactical Communications 2009 (Situational Awareness and Operational Effectiveness in the Last Tactical Mile), One Whitehall Place, Whitehall Suite and Reception, London, UK (2009)
101. Sapaty, P.S.: Distributed capability for battlespace dominance. In: Electronic Warfare 2009 Conference and Exhibition, Novotel London West Hotel and Conference Center, London (2009)
102. Sapaty, P.S.: Providing spatial integrity for distributed unmanned systems. In: Proceedings of 6th International Conference in Control, Automation and Robotics ICINCO 2009, Milan, Italy (2009)
103. Sapaty, P.S.: Distributed technology for global control. In: Lecture Notes in Electrical Engineering, Informatics in Control, Automation and Robotics, vol. 37. Springer, Berlin (2009)
104. Sapaty, P.S.: Gestalt-based integrity of distributed networked systems. In: SPIE Europe Security + Defence, bcc Berliner Congress Centre, Berlin, Germany (2009)
105. Sapaty, P.S.: Remote control of open groups of remote sensors. In: Proceedings of SPIE Europe Security + Defence, Berlin, Germany (2009)
106. Sapaty, P.S.: Tactical communications in advanced systems for asymmetric operations. In: Proceedings of Tactical Communications 2010, CCT Venues, Canary Wharf, London UK (2010)

107. Sapaty, P.S.: High-level technology to manage distributed robotized systems. In: Proceedings of Military Robotics 2010. Jolly St. Ermins, London, UK (2010)
108. Sapaty, P.S.: Emerging asymmetric threats, Q&A session. In: Tactical Communications 2010, CCT Venues. Canary Wharf, London UK (2010)
109. Sapaty, P.S.: High-level organization and management of directed energy systems. In: Proceedings of Directed Energy Weapons 2010, CCT. Canary Wharf, London UK (2010)
110. Sapaty, P.S.: Unified transition to robotized armies with spatial grasp technology. In: International Summit Military Robotics, London, United Kingdom (2012)
111. Sapaty, P.S.: Distributed air and missile defense with spatial grasp technology. Intell. Control Autom. Sci. Res. **3**(2), 117–131 (2012)
112. Sapaty, P.S.: Global electronic dominance. In: 12th International Fighter Symposium, Grand Connaught Rooms, London, UK (2012)
113. Sapaty, P.S.: Providing global awareness in distributed dynamic environments. In: International Summit ISR, London (2013)
114. Sapaty, P.S.: Ruling distributed dynamic worlds with spatial grasp technology. In: Tutorial at the International Science and Information Conference 2013 (SAI), London, UK (2013)
115. Sapaty, P.S.: Night vision under advanced spatial intelligence: a key to battlefield dominance. In: International Summit Night Vision 2013, London (2013)
116. Sapaty, P.S.: Integration of ISR with advanced command and control for critical mission applications. In: SMi's ISR Conference, Holiday Inn Regents Park, London (2014)
117. Sapaty, P.S.: Unified transition to cooperative unmanned systems under spatial grasp paradigm. Int. J. Trans. Netw. Commun. **2**(2) (2014)
118. Sapaty, P.S.: From manned to smart unmanned systems: a unified transition. In: SMi's Military Robotics, Holiday Inn Regents Park London (2014)
119. Sapaty, P.S.: Unified transition to cooperative unmanned systems under spatial grasp paradigm. In: 19th International Command and Control Research and Technology Symposium, Alexandria, Virginia (2014)
120. Sapaty, P.S.: Distributed human terrain operations for solving national and international problems. Int. Relat. Dipl. **2**(9) (2014)
121. Sapaty, P.S.: Providing over-operability of advanced ISR systems by a high-level networking technology. In: SMI's Airborne ISR, Holiday Inn Kensington Forum, London, United Kingdom (2015)
122. Sapaty, P.S.: Distributed missile defence with spatial grasp technology. In: SMi's Military Space, Holiday Inn Regents Park London (2015)
123. Sapaty, P.S.: Military robotics: latest trends and spatial grasp solutions. Int. J. Adv. Res. Artif. Intell. **4**(4) (2015)
124. Sapaty, P.S., A brief introduction to the spatial grasp language (SGL). J. Comput. Sci. Syst. Biol. **09**(02) (2015). https://www.researchgate.net/publication/307744279_A_Brief_Introduction_to_the_Spatial_Grasp_Language_SGL
125. Sapaty, P.S.: Organization of advanced ISR systems by high-level networking technology. Math. Mach. Syst. (1) (2016)
126. Sapaty, P.S.: Towards massively robotized systems under spatial grasp technology. J. Comput. Sci. Syst. Biol. **9**(1) (2016)
127. Sapaty, P.S.: Towards wholeness and integrity of distributed dynamic systems. J. Comput. Sci. Syst. Biol. **9**(3) (2016)
128. Sapaty, P.S.: Towards global goal orientation, robustness and integrity of distributed dynamic systems. J. Int. Relat. Dipl. **4**(6) (2016)
129. Sapaty, P.: Spatial Grasp Language (SGL). Adv. Image Video Process. **4**(1) (2016). https://journals.scholarpublishing.org/index.php/AIVP/article/view/1922/1052
130. Sapaty P.S.: Mosaic warfare: from philosohpy to model to solution. Math. Mach. Syst. (3) (2019)
131. Sapaty, P.S.: Global network management under spatial grasp paradigm. Int. Robot. Autom. J. **6**(3), 134–148 (2020). https://medcraveonline.com/IRATJ/IRATJ-06-00212.pdf

132. Sapaty, P.S.: Global network management under spatial grasp paradigm. Glob. J. Res. Eng.: J. Gener. Eng. **20**(5), 58–81 (Version 1.0) (2020). https://globaljournals.org/GJRE_Volume20/6-Global-Network-Management.pdf
133. Sapaty, P.: Advanced terrestrial and celestial missions under spatial grasp technology. Aeron. Aerosp. Open Access J. **4**(3) (2020). https://medcraveonline.com/AAOAJ/AAOAJ-04-00110.pdf
134. Sapaty, P.: Spatial management of distributed social systems. J. Comput. Sci. Res. **02**(03) (2020). https://ojs.bilpublishing.com/index.php/jcsr/article/view/2077/pdf
135. Sapaty, P.: Towards global nanosystems under high-level networking technology. Acta Sci. Comput. Sci. **2**(8) (2020). https://www.actascientific.com/ASCS/pdf/ASCS-02-0051.pdf
136. Sapaty, P.: Symbiosis of distributed simulation and control under spatial grasp technology. SSRG Int. J. Mob. Comput. Appl. (IJMCA) **7**(2) (2020). http://www.internationaljournalssrg.org/IJMCA/2020/Volume7-Issue2/IJMCA-V7I2P101.pdf
137. Sapaty, P.S.: Symbiosis of real and simulated worlds under global awareness and consciousness. In: Abstract, The Science of Consciousness (TSC) conferences, Tucson, Arizona (2020). https://eagle.sbs.arizona.edu/sc/report_poster_detail.php?abs=3696
138. Sapaty, P.S.: Spatial Grasp as a model for space-based control and management systems. Math. Mach. Syst. (1), 135–138 (2021). http://www.immsp.kiev.ua/publications/articles/2021/2021_1/Sapaty_book_1_2021.pdf
139. Sapaty, P.S.: Managing multiple satellite architectures by spatial grasp technology. Math. Mach. Syst. (1), 3–16 (2021). http://www.immsp.kiev.ua/publications/eng/2021_1/
140. Sapaty, P.S.: Spatial management of large constellations of small satellites. Math. Mach. Syst. (2) (2021). http://www.immsp.kiev.ua/publications/articles/2021/2021_2/02_21_Sapaty.pdf
141. Sapaty, P.S.: Global management of space debris removal under spatial grasp technology. Acta Sci. Comput. Sci. **3**(7) (2021). https://www.actascientific.com/ASCS/pdf/ASCS-03-0135.pdf
142. Sapaty, P.S.: Space debris removal under spatial grasp technology. Netw. Commun. Technol. **6**(1) (2021). https://www.ccsenet.org/journal/index.php/nct/article/view/0/45486
143. Sapaty, P.S.: Relation of spatial grasp paradigm to higher psychological and mental concepts. Acta Sci. Comput. Sci. **4**(12) (2022). https://actascientific.com/ASCS/pdf/ASCS-04-0359.pdf
144. Sapaty, P.S.: Seeing and managing distributed worlds with spatial grasp paradigm. Acta Sci. Comput. Sci. **4**(12) (2022). https://actascientific.com/ASCS/pdf/ASCS-04-0365.pdf
145. Sapaty, P.S.: Comprehending distributed worlds with the spatial grasp paradigm. Math. Mach. Syst. (1) (2022). http://www.immsp.kiev.ua/publications/articles/2022/2022_1/01_22_Sapaty.pdf

Chapter 3
Spatial Grasp Model and Technology Basics

3.1 Introduction

Having rich history and experience of distributed computer networking, this recursive super-virus model and technology, which is self-evolving, self-migrating, and self-matching in distributed environments, is briefed and explained. It can dynamically establish and keep superior vision and power over any distributed systems, at the same time being independent of their organizations and networked structures but rather creating, supervising and modifying them if needed.

The rest of the chapter is organized as follows. Section 3.2 describes Spatial Grasp Technology general issues. Section 3.3 provides main Spatial Grasp Language features including its recursive structure, constants, variables, rules, top level summary, and how SGL scenarios evolve with their control states. Section 3.4 explains details of networked SGL implementation which include key language interpretation features, SGL interpreter networked organization allowing it operate as a powerful spatial engine, also its spatial tracking system effectively supporting distributed management and control. Section 3.5 compares SGT with traditional ideology of mobile agents based on quite different and inferior principles. Section 3.6 concludes the chapter. Selected References reflect some of many existing publications on this spatial paradigm, covering European Patent, previous books, latest journal papers, and earlier technology versions including those called WAVE.

3.2 General Technology Issues

Within ***Spatial Grasp Model and Technology (SGT)*** (see the related patent [1], published books [2–10], latest journal papers [11–18], many previous publications on this paradigm, only a few by [19–81]), a high-level operational scenario expressed

P. S. Sapaty, *Spatial Networking in the United Physical, Virtual, and Mental World*, Studies in Systems, Decision and Control 533,
https://doi.org/10.1007/978-3-031-62154-3_3

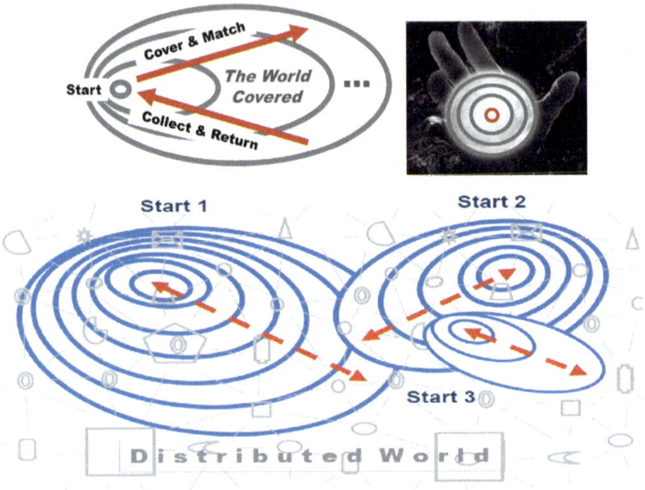

Fig. 3.1 Symbolic views of parallel recursive world coverage with Spatial Grasp model

in recursive ***Spatial Grasp Language (SGL)***, starting in any world point *propagates, covers, and matches the distributed environment in parallel wavelike mode*, as symbolically shown from different views in Fig. 3.1. Such propagation can result in returning and analyzing the reached states and data, which may be arbitrarily remote, also used for launching further waves from them or start again, jointly from the both too.

This concept is based on quite different philosophy of dealing with large distributed systems. Instead of representing systems and solutions in them in the form of communicating parts or agents, the developed Spatial Grasp paradigm is organizing everything by the *integral, holistic and parallel substance* (like flood, smog, even virus) covering and conquering distributed worlds which may be as follows:

Physical World (PW), considered as continuous and infinite where each point can be identified and accessed by physical coordinates,

Virtual World (VW), which is discrete and consists of nodes and semantic links between them,

Executive world (EW), consisting of active "doers", which may be humans or robots, with communication possibilities between them.

Different kinds of combinations of these worlds can also be possible within the same formalism, like Virtual-Physical World (VPW), Virtual-Execution World (VEW), Execution-Physical World (EPW), and Virtual-Execution-Physical World (VEPW) combining all features of the previous cases.

3.3 Spatial Grasp Language Features

- **SGL recursive structure**

The language top level recursive organization can be expressed just by a single string, in a formula-like mode, as follows:

$$\boldsymbol{grasp \rightarrow constant|variable|rule(\{grasp,\ \})}$$

An SGL scenario, called ***grasp***, applied in some point of the distributed space, can just be a ***constant*** directly providing the result to be associated with this point. It can be a ***variable*** whose content, assigned to it previously when staying in this or another space point (as variables may have non-local meaning and coverage), provides the result in the application point too. It can also be a ***rule*** (expressing certain ordering, action, control, description, or context) optionally accompanied with operands separated by comma and embraced in parentheses. These operands can be of any nature and complexity (including arbitrary scenarios themselves) and defined recursively as *grasp* again, i.e. can be *constants*, *variables*, or any *rules* with operands.

Some information on constants, variables and rules as follows.

- **SGL constants**

Constants are self-identifiable by the way they are written. They can also be explicitly defined by special embracing rules for arbitrary textual representations. They may be:

Information—As arbitrary strings in single quotes (which may be nested, if multiple), also numbers in traditional representations;

Physical matter or physical objects—As any texts in double quotes (which can be nested too) identifying adequately these elements;

Special or reserved constants—which may be used as standard parameters or modifiers in different language rules.

Custom items—Special self-identifiable names/words established additionally by customers to serve specific classes of problems (to be adjusted with SGL implementations).

Details on different constant types and their usage are in Chap. 4.

- **SGL variables**

Their repertoire is as follows, all being self-identifiable or declared by special rules for any names:

Global variables (most expensive and rarely used), which can serve any SGL scenarios and be shared by them,

Heritable variables appearing within a scenario step and serving all subsequent steps,

Frontal variables serving and accompanying the scenario evolution, being transferred between subsequent steps,

Nodal variables as a set up property of the world positions reached by scenarios and shared with other scenarios in the same positions,

Environmental variables allowing us to access, analyze, and change different parameters of physical, virtual and executive worlds during their navigation, also certain language interpretation features (as of the internal implementation world).

- **SGL rules**

SGL rules, starting in world points, can organize navigation of the world sequentially, in parallel, or any combinations thereof, with their types being as: ***type***, ***usage***, ***seeing***, ***movement***, ***creation***, ***echoing***, ***verification***, ***assignment***, ***advancement***, ***branching***, ***transference***, ***exchange***, ***timing***, *and* ***qualifying***.

The rules can result in staying in the same application point or cause movement to other world points with the obtained results to be left there or returned. The points reached can become starting points for other rules. The rules, due to recursive language organization, can form *arbitrary operational and control infrastructures* expressing any sequential, parallel, hierarchical, centralized, localized, dissipated, mixed, and up to fully decentralized and distributed algorithms. The concept of *rule* is dominant in SGL not only for diverse activities on data, knowledge, and physical matter, but also for overall management and control of any SGL scenarios.

- **SGL top level summary**

Top SGL recursive organization can be expressed as:

Grasp → *constant* | *variable* | *rule* ({ *grasp*, })

constant → *information* | *matter* | *custom* | *special* | *grasp*

variable → *global* | *heritable* | *frontal* | *nodal* | *environmental* | *grasp*

rule → *type* | *usage* | *movement* | *creation* | *echoing* |

verification | *assignment* | *advancement* | *branching* |

transference | *exchange* | *timing* | *qualifying* | *grasp*

Its recursive syntax can also be expressed graphically, as in Fig. 3.2.

As can be seen from this description, the basic language constructs like ***constants***, ***variables***, and ***rules*** can themselves be defined as ***grasps*** too, i.e. represented by individual scenarios of any complexity and space coverage. These can potentially result in data structures of any complexity, names or collection of names of variables to be processed together, and names of rules or their collections that should (individually or jointly) manage the embraced scenarios.

- **How SGL scenarios evolve**

The following are some hints on how SGL scenarios self-evolve in distributed environments.

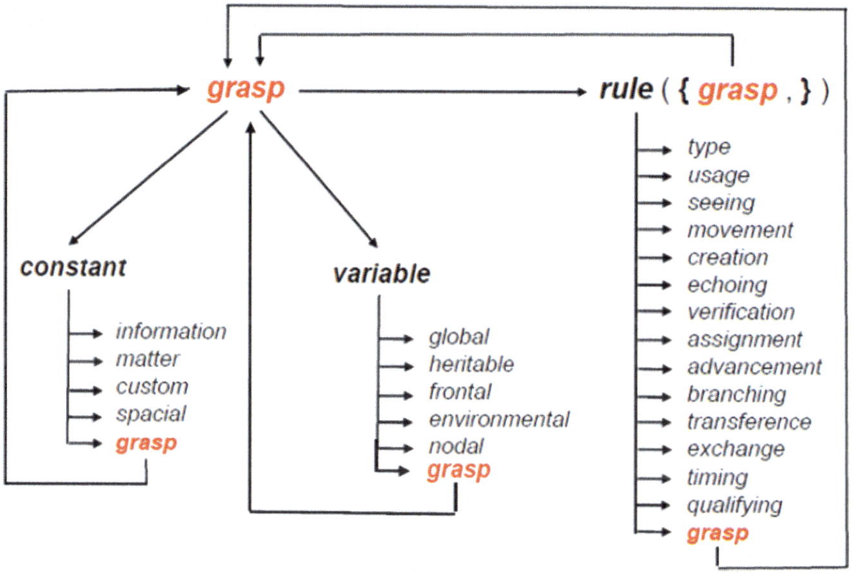

Fig. 3.2 SGL top level recursive organization

- SGL scenario is developing in *steps*, potentially parallel, with new steps produced on the results of previous steps.
- Any step is always associated with a certain *point* or points of the world (physical, virtual, executive, or combined) from which the scenario is currently developing.
- Each step provides a resultant *value* (single or multiple) representing information, matter, or both, and a resultant control *state*.
- Different scenario parts may evolve from the same points in *ordered*, *unordered*, or *parallel* manner, as independent or interdependent branches.
- Different scenario parts can spatially *succeed* each other, with new parts evolving from final positions reached by the previous parts.
- This potentially parallel and distributed scenario may proceed in *synchronous* or *asynchronous* modes, also their any combinations.
- SGL operations and decisions can use *control states* and *values* returned from other, subsequent, scenario parts, combining forward and backward scenario evolution.
- Different steps from the same or different scenarios may be temporarily associated with *same world points,* sharing persistent or provisional information in them.
- Staying with different world points, whether physical or virtual, it may be possible to *change local parameters* in them, thus impacting the worlds via these locations.
- Scenarios navigating distributed spaces can create active *distributed physical or virtual infrastructures* in them operating on their own, also shared with other scenarios.

- Overall organization of the world creation, coverage, modification, analysis and processing can be provided by a variety of SGL rules which may be *arbitrarily nested*.
- The evolving SGL scenarios can *lose utilized parts* if not needed any more, also *self-modify* and *self-replicate* during space navigation, automatically adjusting to the distributed environments.

- **SGL control states**

The following control states appear after completion of different scenario steps, indicating their progress or failure. They can be used for effective control of multiple distributed processes with proper decisions at different levels.

Thru—reflects full success of the current branch with capability of further scenario development;

done—indicates success of the current scenario step with its planned termination;

fail—indicates non-revocable failure of the current branch and no possibility of further development from the location reached;

fatal—reporting terminal failure in the location reached while triggering massive abortion of the currently evolving scenario processes, in other branches too, with deletion of associated temporary data in them. The level of such removal process can be controlled by special higher-level rules.

Full details of SGL are provided in the next chapter.

3.4 Networked SGL Implementation

- **General features**

The ***SGL interpreter*** consists of specialized modules working with specific data structures, serving SGL scenarios or their parts happened to be inside this interpreter, also organizing exchanges with other interpreters for distributed SGL scenarios. Each interpreter copy can process multiple active scenario code propagating in space and time. Communicating SGL interpreters can be in *arbitrary number of copies* effectively integrated with other existing systems and communications, altogether representing powerful spatial engines operating without central resources or control. Hardware or software SGL interpreters, shown in Fig. 3.3 as universal control and processing units *effectively working with spatial graph and network data*, can be installed, also runtime created, in proper physical or virtual world points.

- **General SGL interpreter organization**

The interpreter copy consists of a number of specialized functional modules, as in Fig. 3.4 and mentioned below, working with specific data structures, both serving

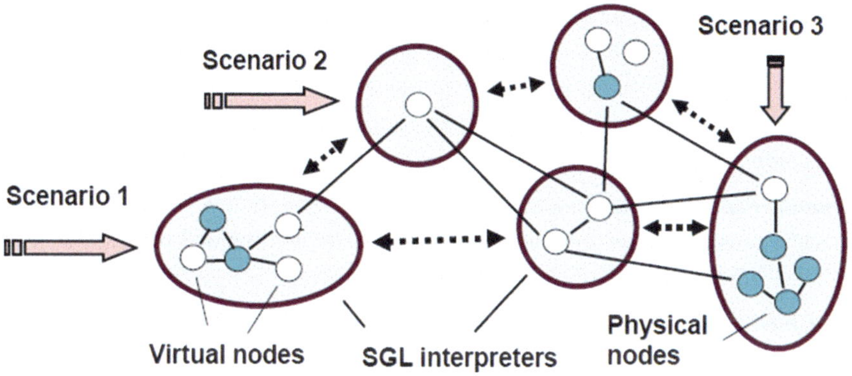

Fig. 3.3 SGL distributed interpretation processing complex networked data

SGL scenarios or their parts which happen to be inside this interpreter at this moment of time, also organizing exchanges with other interpreters in case of distributed SGL scenarios.

Grasps queue management—serving multiple SGL scenarios currently staying partially or entirely within the current interpreter copy, taking into account their priorities or just serving in a FIFO manner.

Data processing—executes different logical, numerical and string operations activated by SGL rules, which may also include operations on physical matter and those providing access to external environment.

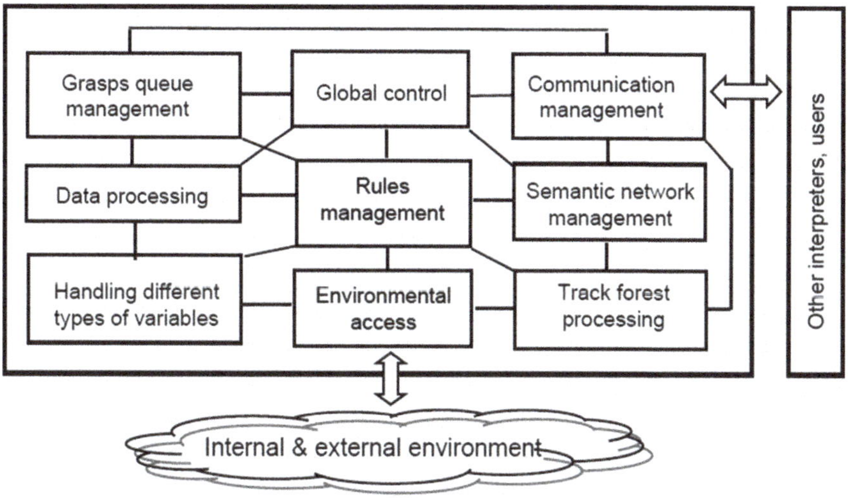

Fig. 3.4 SGL interpreter main components and their interactions

Handling different types of variables. Keeps global, nodal, heritable and environmental variables currently associated with the covered parts of virtual and physical spaces, also temporarily serving frontal variables before forwarding them to other interpreters.

Rules management—holds and implements semantics of different rules.

Semantic network management—keeps and processes part of the semantic network currently covered by the current interpreter, also supporting and processing links with its parts in other interpreters.

Track forest processing—organizes creation, support and modification of distributed control tracks during spatial development of parallel scenarios, integrates with track forests in other interpreters, also supports spatial variables (some being mobile) and forward–backward control propagation via the tracks.

Communication management—organizes and supports communications with other interpreters and users within integral spatial SGL solutions.

Environmental access—organizes and implements communications with external and internal virtual and physical environments based on the environmental variables.

Global control—supports integrity of multiple data processing, communications, and management solutions within the current interpreter (including timing of critical operations), also coordinating with global control in other interpreters to provide integrity of spatial solutions, which may have any terrestrial or celestial nature.

- **Interpretation network as a spatial engine**

Communicating interpreters of SGL can be in arbitrary number of copies, up to millions and billions, which can be effectively integrated with any existing systems and communications, and their dynamic networks can represent *powerful spatial engines capable of solving any problems in terrestrial and celestial environments*. Such collective engines can simultaneously execute different cooperative or competitive scenarios without any central resources or control. Hardware or software SGL interpreters may be stationary or mobile. They can be dynamically installed, and if needed created, in proper physical or virtual world points on the request of self-evolving SGL scenarios.

- **Spatial tracking system**

As *both backbone and nerve system of the distributed interpreter*, its self-optimizing *Spatial Track System* provides hierarchical command and control, as well as remote data and code access, It also supports spatial variables and merges distributed control states for decisions at higher organizational levels. The track infrastructure is automatically distributed between active components (humans, robots, computers, smartphones, satellites, etc.) during scenario self-spreading in distributed environments. It integrates the following stages of operation (some stages of track system operation symbolically shown in Fig. 3.5).

Some more details for the track operations.

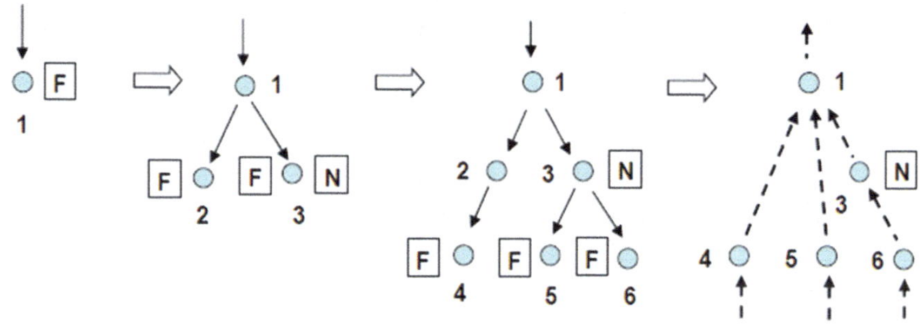

Fig. 3.5 Different spatial track system operations. **a** Start with a single track node and frontal variable; **b** Track grows with replication of frontal variable and fixed support of nodal variable; **c** Further track growing with movement and replication of frontal variables; **d** Echoing via tracks with optimization of tracks structure and continuing support of the nodal variable

Forward grasping. In the forward process, next steps of scenario development are forming new track nodes connected to the previous nodes by track links. Reflecting the history of scenario evolution, this growing track structure is effectively supporting heritable, nodal and frontal variables associated with track nodes.

Echoing. After completing the forward stage of SGL scenario, the track system can return to the starting track node the generalized control state based on termination states in all fringe nodes, also marking the passed track links with the states returned via them. The track system, on the request of higher-level scenario rules, can also collect the local data obtained at its fringe nodes and process them hierarchically and in parallel, or just merge them into a resultant list of values, to be echoed to the higher levels. The track echoing process also optimizes the track system by removing already used and not needed any more items.

Further forward development. The echo-modified and optimized track system can route further grasps to the world positions reached by the previous grasps—to all or only part of them, as regards the higher-level decisions. Heritable variables created in certain track nodes can be also accessed from the subsequent nodes in the track system for both reading and writing operations.

Much more detailed SGL interpreter structure, organization and operation are described in Chap. 5.

3.5 Comparison of SGT with Mobile Agents

A mobile agent [82–87] is a piece of software, combined with data, capable to migrate from one computer to another autonomously and continue its execution on the destination. Mobile agent is a process that can transport its state from one environment to another, and be capable of performing in the new environment. In the early 1990s, the UScompany General Magic created Telescript language [87] and environment for writing and executing mobile agents. The briefed above Spatial Grasp Model and Technology is also based on code mobility in computer networks, but assigns this global feature to the whole programming system in which self-replicating parts of spatial scenarios can propagate in parallel in distributed environments and communicate under recursive spatial control, also mobile.

SGT keeps all functionality and power within self-replicating virus-like code with unlimited power to create, modify, control and manage any distributed systems, actually existing independently over them. This paradigm (originally called WAVE) appeared much earlier than mobile agents and Telescriprt (actually from the end of the sixties), and was practically used to organize distributed computations in city-wide heterogeneous computer networks [19–29]. The current author visited General Magic in Seattle in the past with presentation of WAVE technology there (which was installed and operating at that time at UBC in Vancouver, Canada).

3.6 Conclusions

General ideas of the mobile Spatial Grasp Technology (SGT) and its basic Spatial Grasp Language (SGL) have been summarized. The organization of SGL interpreter was briefed, with its communicating copies forming powerful parallel networked engines covering arbitrary large environments and effectively working with distributed networking models and systems. The purpose and organization of the spatial tracking system of the interpreter effectively supporting spatial forward–backward capabilities of SGL have been explained. Comparison of SGT with mobile agents provided too.

References

1. Sapaty, P.S.: A distributed processing system, European Patent N 0389655, Publ. 10.11.93, European Patent Office
2. Sapaty, P.S.: Mobile Processing in Distributed and Open Environments. John Wiley & Sons, New York (1999)
3. Sapaty, P.S.: Ruling Distributed Dynamic Worlds. John Wiley & Sons, New York (2005)
4. Sapaty, P.S.: Managing Distributed Dynamic Systems with Spatial Grasp Technology. Springer (2017)
5. Sapaty, P.S.: Holistic Analysis and Management of Distributed Social Systems. Springer (2018)

6. Sapaty, P.S.: Complexity in International Security: A Holistic Spatial Approach. Emerald Publishing (2019)
7. Sapaty, P.S.: Symbiosis of Real and Simulated Worlds under Spatial Grasp Technology. Springer (2021)
8. Sapaty, P.S.: Spatial Grasp as a Model for Space-based Control and Management Systems. CRC Press (2022)
9. Sapaty, P.S.: The Spatial Grasp Model: Applications and Investigations of Distributed Dynamic Worlds. Emerald Points (2023)
10. Sapaty, P.S.: Providing Integrity, Awareness, and Consciousness in Distributed Dynamic Systems. CRC Press (2024)
11. Sapaty P.S.: Spatial management of air and missile defence operations. Math. Mach. Syst **1** (2023). http://www.immsp.kiev.ua/publications/articles/2023/2023_1/01_23_Sapaty.pdf
12. Sapaty P.S.: Providing distributed system integrity under Spatial Grasp Technology. Math. Mach. Syst. **2** (2023). http://www.immsp.kiev.ua/publications/articles/2023/2023_2/02_23_Sapaty.pdf
13. Sapaty P.S.: Providing global awareness in distributed dynamic systems. Int. Relat. Diplomacy **11**(2), 87–100 (2023). https://doi.org/10.17265/2328-2134/2023.02.002. https://www.davidpublisher.com/Public/uploads/Contribute/6486c3d05a6cc.pdf
14. Sapaty, P.S.: Simulating distributed consciousness with Spatial Grasp Model. Math. Mach. Syst. **3** (2023)
15. Sapaty P.S.: Network centrality operations under Spatial Grasp Technology. J. Adv. Artif. Intell. Mach. Learn. **1**(1), 1–11 (2023). https://www.scivisionpub.com/journals/articleinpress-journal-of-advances-in-artificial-intelligence-and-machine-learning
16. Sapaty, P.S.: Managing distributed systems with Spatial Grasp patterns. Math. Mach. Syst. **4** (2023). http://www.immsp.kiev.ua/publications/articles/2023/2023_4/04_23_Sapaty.pdf
17. Sapaty, P.S.: Spatial networks as models for organoid cultures and brain research. Math. Mach. Syst. **1** (2024)
18. Sapaty, P.S.: Basic network operations under Spatial Grasp Model. Sci. Set J. Phys. **3**(1), 1–16 (2024)
19. Bondarenko, A.T., Mikhalevich, S.B., Nikitin, A.I., Sapaty, P.S.: Software of BESM-6 computer for communication with peripheral computers via telephone channels. In: Computer Software, Vol. 5, Inst. of Cybernetics Press, Kiev (1970)
20. Sapaty, P.S.: A method of organization of an intercomputer dialogue in the radial computer systems. In: The Design of Software and Hardware for Automatic Control Systems. Institute of Cybernetics Press, Kiev (1973)
21. Sapaty, S.: A wave language for parallel processing of semantic networks. Comput. Artif. Intell. **5**(4), 289–314 (1986)
22. Sapaty, P.S.: The WAVE-0 language as a framework of navigational structures for knowledge bases using semantic networks. In: Proceedings of USSR Academy of Sciences. Technical Cybernetics, No. 5 (1986)
23. Sapaty, P.S.: The WAVE-1: A new ideology and language of distributed processing on graphs and networks. Comput. Artif. Intell. **5** (1987)
24. Sapaty, P.S.: The WAVE model for advanced knowledge processing, Report No. OUEL 1803/89. University of Oxford, England (1989)
25. Sapaty, P.S.: The WAVE Model for advanced knowledge processing. In: Ambler, A.P., Agrawal, P., Moore, W.R. (eds) CAD Accelerators. Elsevier Science Publ. B.V. (1990)
26. Sapaty, P.S.: The WAVE machine project. In: Proceedings of IFIP Workshop on Silicon Architectures for Neural Nets, St. Paul de Vence, France, November 28–30 (1990)
27. Sapaty, P.S.: The WAVE paradigm, Report No. 17/92, Faculty of Informatics, University of Karlsruhe (1992)
28. Sapaty, P.S.: The WAVE paradigm. In: Proceedings of JICSLP'92 Post-Conference Joint Workshop on Distributed and Parallel Implementations of Logic Programming Systems, Washington, D.C., Nov 13–14 (1992)

29. Sapaty, P.S.: Distributed modeling of cooperative behavior by mobile agents. In: Proceedings of Sixth Conference on Computer Generated Forces and Behavioral Representation, IST UCF, Orlando, FL, July, pp. 599–613 (1996)
30. Sapaty, P.S.: Distributed artificial brain for collectively behaving mobile robots. In: Proceedings of Symposium and Exhibition Unmanned Systems 2001, Baltimore, MD, Jul 31–Aug 2 (2001)
31. Sapaty, P.S.: Crisis management with distributed processing technology. Int. Trans. Syst. Sci. Appl. **1**(1), 81–92. ISSN 1751-1461 (2006)
32. Sapaty, P.S.: Global management of distributed EW-related system. In: Proceedings of International Conference Electronic Warfare: Operations & Systems, Thistle (2007)
33. Sapaty, P.S.: Distributed technology for global dominance. In: Proceedings of International Conference Defense Transformation and Net-Centric Systems 2008, as part of the SPIE Defense and Security Symposium, World Center Marriott Resort and Convention Center, Orlando, FL, USA, 16–20 Mar (2008) (Also: Proceedings of SPIE—Volume 6981, Defense Transformation and Net-Centric Systems 2008, Raja Suresh, Editor, 69810T, Apr. 3, 2008)
34. Sapaty, P.S.: Distributed technology for global dominance. Keynote lecture, Proceedings of the Fifth International Conference in Control, Automation and Robotics ICINCO 2008, The Conference Proceedings, Funchal, Madeira, Portugal, 11–15 May (2008)
35. Sapaty, P.S.: Human-robotic teaming: a compromised solution. In: AUVSI's Unmanned Systems North America 2008, San Diego, USA, 10–12 June (2008)
36. Sapaty, P.S.: High-level communication protocol for dynamically networked battlefields. In: Proceedings of International Conference Tactical Communications 2009 (Situational Awareness and Operational Effectiveness in the Last Tactical Mile), One Whitehall Place, Whitehall Suite and Reception, London, UK (2009)
37. Sapaty, P.S.: Distributed capability for battlespace dominance. In: Electronic Warfare 2009 Conference and Exhibition, Novotel London West Hotel and Conference Center, London, 14–15 May (2009)
38. Sapaty, P.S.: Providing spatial integrity for distributed unmanned systems. In: Proceedings of 6th International Conference in Control, Automation and Robotics ICINCO 2009, Milan, Italy (2009)
39. Sapaty, P.S.: Distributed technology for global control. In: Book chapter, Lecture Notes in Electrical Engineering, Informatics in Control, Automation and Robotics, vol. 37. Springer, Berlin (2009)
40. Sapaty, P.S.: Gestalt-based integrity of distributed networked systems. In: SPIE Europe Security + Defence, bcc Berliner Congress Centre, Berlin, Germany (2009)
41. Sapaty, P.S.: Remote control of open groups of remote sensors. In: Proceedings of SPIE Europe Security + Defence, Berlin, Germany (2009)
42. Sapaty, P.S.: Tactical communications in advanced systems for asymmetric operations. In: Proceedings of Tactical Communications 2010, CCT Venues, Canary Wharf, London UK, 28–30 April (2010)
43. Sapaty, P.S.: High-level technology to manage distributed robotized systems. In: Proceedings of Military Robotics 2010, Jolly St Ermins, London, UK, 25–27 May (2010)
44. Sapaty, P.S.: Emerging asymmetric threats, Q&A Session. Tactical Communications 2010, CCT Venues, Canary Wharf, London UK, 28–30 Apr (2010)
45. Sapaty, P.S.: High-level organization and management of directed energy systems. In: Proceedings of Directed Energy Weapons 2010, CCT, Canary Wharf, London UK, 25–26, Mar (2010)
46. Sapaty, P.S.: Unified transition to robotized armies with Spatial Grasp Technology. International Summit Military Robotics, London, United Kingdom, 12–13 Nov (2012)
47. Sapaty, P.S.: Distributed air and missile defense with Spatial Grasp Technology. Intell. Control Autom. Sci. Res. **3**(2), 117–131 (2012)
48. Sapaty, P.S.: Global electronic dominance. In: 12th International Fighter Symposium, 6th–8th November 2012, Grand Connaught Rooms, London, UK
49. Sapaty, P.S.: Providing global awareness in distributed dynamic environments. In: International summit ISR, London, 16–18 Apr (2013)

50. Sapaty, P.S.: Ruling distributed dynamic worlds with Spatial Grasp Technology. In: Tutorial at the International Science and Information Conference 2013 (SAI), London, UK, 7–9 Oct (2013)
51. Sapaty, P.S.: Night vision under advanced spatial intelligence: a key to battlefield dominance. In: International Summit Night Vision 2013, London, 4–6 June (2013)
52. Sapaty, P.S.: Integration of ISR with advanced command and control for critical mission applications. In: SMi's ISR Conference, Holiday Inn Regents Park, London, 7–8 Apr (2014)
53. Sapaty, P.S.: Unified transition to cooperative unmanned systems under Spatial Grasp paradigm. Int. J. Trans. Netw. Commun. **2**(2) (2014)
54. Sapaty, P.S.: From manned to smart unmanned systems: a unified transition. In: SMi's Military Robotics. Holiday Inn Regents Park London, 21–22 May (2014)
55. Sapaty, P.S.: Unified transition to cooperative unmanned systems under Spatial Grasp paradigm. In: 19th international command and control research and technology symposium, Alexandria, Virginia, 16–19 June (2014)
56. Sapaty, P.S.: Distributed human terrain operations for solving national and international problems. Int. Relat. Dipl. **2**(9) (2014)
57. Sapaty, P.S.: Providing over-operability of advanced ISR systems by a high-level networking technology. In: SMI's Airborne ISR, Holiday Inn Kensington Forum, London, United Kingdom, 26–27 Oct (2015)
58. Sapaty, P.S.: Distributed missile defence with Spatial Grasp Technology. In: SMi's Military Space, Holiday Inn Regents Park London, 4–5 Mar (2015)
59. Sapaty, P.S.: Military robotics: latest trends and Spatial Grasp solutions. Int. J. Adv. Res. Artif. Intell. **4**(4) (2015)
60. Sapaty, P.S.: A Brief Introduction to the Spatial Grasp Language (SGL). J. Comput. Sci. Syst. Biol. **9**(2) (2015). https://www.researchgate.net/publication/307744279_A_Brief_Introduction_to_the_Spatial_Grasp_Language_SGL
61. Sapaty, P.S.: Organization of advanced ISR systems by high-level networking technology. Math. Mach. Syst. **1** (2016)
62. Sapaty, P.S.: Towards massively robotized systems under Spatial Grasp Technology. J. Comput. Sci. Syst. Biol. **9**(1) (2016)
63. Sapaty, P.S.: Towards wholeness and integrity of distributed dynamic systems. J. Comput. Sci. Syst. Biol. **9**(3) (2016)
64. Sapaty, P.S.: Towards global goal orientation, robustness and integrity of distributed dynamic systems. J. Int. Relat. Dipl. **4**(6) (2016)
65. Sapaty, P.: Spatial Grasp Language (SGL). Adv. Image Video Process. **4**(1) (2016). https://journals.scholarpublishing.org/index.php/AIVP/article/view/1922/1052
66. Sapaty, P.S.: Mosaic warfare: from philosohpy to model to solution. Math. Mach. Syst. **3** (2019)
67. Sapaty, P.S.: Global network management under Spatial Grasp paradigm. Int. Robot. Autom. J. **6**(3), 134–148 (2020). https://medcraveonline.com/IRATJ/IRATJ-06-00212.pdf
68. Sapaty, P.S.: Global network management under Spatial Grasp paradigm. Global J. Res. Eng.: J General Eng. **20**(5), 58–81 (2020). Version 1.0. https://globaljournals.org/GJRE_Volume20/6-Global-Network-Management.pdf
69. Sapaty, P.: Advanced terrestrial and celestial missions under Spatial Grasp Technology. Aeronaut. Aerosp. Open Access J. **4**(3) (2020). https://medcraveonline.com/AAOAJ/AAOAJ-04-00110.pdf
70. Sapaty, P.: Spatial management of distributed social systems. J. Comput. Sci. Res. **2**(3) (2020). https://ojs.bilpublishing.com/index.php/jcsr/article/view/2077/pdf
71. Sapaty, P.: Towards global nanosystems under high-level networking technology. Acta Sci. Comput. Sci. **2**(8) (2020). https://www.actascientific.com/ASCS/pdf/ASCS-02-0051.pdf
72. Sapaty, P.: Symbiosis of distributed simulation and control under Spatial Grasp Technology. SSRG Int. J. Mobile Comput. Appl. (IJMCA) **7**(2) (2020). http://www.internationaljournalssrg.org/IJMCA/2020/Volume7-Issue2/IJMCA-V7I2P101.pdf
73. Sapaty, P.S.: Symbiosis of real and simulated worlds under global awareness and consciousness. Abstract, The Science of Consciousness (TSC) Conferences, Tucson, Arizona (2020). https://eagle.sbs.arizona.edu/sc/report_poster_detail.php?abs=3696

74. Sapaty, P.S.: Spatial Grasp as a model for space-based control and management systems. Math. Mach. Syst. **1**, 135–138 (2021). http://www.immsp.kiev.ua/publications/articles/2021/2021_1/Sapaty_book_1_2021.pdf
75. Sapaty, P.S.: Managing multiple satellite architectures by Spatial Grasp Technology. Math. Mach. Syst. **1**, 3–16 (2021). http://www.immsp.kiev.ua/publications/eng/2021_1/
76. Sapaty, P.S.: Spatial management of large constellations of small satellites. Math. Mach. Syst. **2** (2021). http://www.immsp.kiev.ua/publications/articles/2021/2021_2/02_21_Sapaty.pdf
77. Sapaty, P.S.: Global management of space Debris removal under Spatial Grasp Technology. Acta Sci. Comput. Sci. **3**(7) (2021). https://www.actascientific.com/ASCS/pdf/ASCS-03-0135.pdf
78. Sapaty, P.S.: Space debris removal under Spatial Grasp Technology. Netw. Commun. Technol. **6**(1) (2021). https://www.ccsenet.org/journal/index.php/nct/article/view/0/45486
79. Sapaty, P.S.: Relation of Spatial Grasp paradigm to higher psychological and mental concepts. Acta Sci. Comput. Sci. **4**(12) (2022). https://actascientific.com/ASCS/pdf/ASCS-04-0359.pdf
80. Sapaty, P.S.: Seeing and managing distributed worlds with Spatial Grasp paradigm. Acta Sci. Comput. Sci. **4**(12) (2022). https://actascientific.com/ASCS/pdf/ASCS-04-0365.pdf
81. Sapaty, P.S.: Comprehending distributed worlds with the Spatial Grasp paradigm. Math. Mach. Syst. **1** (2022). http://www.immsp.kiev.ua/publications/articles/2022/2022_1/01_22_Sapaty.pdf
82. Mobile agent. https://en.wikipedia.org/wiki/Mobile_agent
83. Agostaro, F., Gencoa, A., Sorce, S.: Mobile agent application fields. Adv. Parall. Comput. **13** (2004). https://www.sciencedirect.com/science/article/abs/pii/S0927545204801102
84. Gray, R., Kotz, D., Nog, S., et al.: Mobile agents: the next generation in distributed computing. In: Proceedings of IEEE International Symposium on Parallel Algorithms Architecture Synthesis, March (1997). https://ieeexplore.ieee.org/document/581620
85. Singh, Y.: Dimensions and Issues of MobileAgent Technology. Int. J. Artif. Intell. Appl. **3**(5), 51–61 (2012). https://www.researchgate.net/publication/276196179_Dimensions_and_Issues_of_Mobile_Agent_Technology
86. Cao, J., Das, S.K.: MobileAgents in Networking and Distributed Computing. Wiley, July (2012). ISBN: 978-0-471-75160-1
87. Telescript Language Reference, General Magic, Inc., October (1995). http://bitsavers.informatik.uni-stuttgart.de/pdf/generalMagic/Telescript_Language_Reference_Oct95.pdf

Chapter 4
Spatial Grasp Language (SGL)

4.1 Introduction

After briefing basics of Spatial Grasp Model and Technology in the previous chapter, where Spatial Grasp Language (SGL) is its main component, the current chapter describes details of SGL constructs and its general organization, which may be particularly helpful for understanding and analysis of numerous practical examples and solutions presented in SGL code in the subsequent chapters. The chapter is organized as follows.

Section 4.2 reminds key features of Spatial Grasp Model and Technology where SGL is its basic language. Section 4.3 provides SGL summary with top language details, language syntax and main constructs. Section 4.4 describes different types of SGL constants which can be self-identifiable by the way they are written or defined by special rules in general case. These include information, physical matter, special constants, compound constants, custom constants, and grasping in most general way. Section 4.5 describes different types of SGL variables with their semantics, spatial distribution, and possible mobility, which include global variables, heritable variables, frontal variables, nodal variables, environmental variables, and grasping as most general case. Section 4.6 lists and explains main SGL rules with their peculiar semantics and practical use, which include: type, usage, movement, creation, echoing, verification, assignment, advancement, branching, transference, exchange, timing, qualification, and grasping as most general case. Section 4.7 provides elementary examples of programming in SGL. Section 4.8 concludes the chapter. References include existing publications on SGT and SGL.

P. S. Sapaty, *Spatial Networking in the United Physical, Virtual, and Mental World*,
Studies in Systems, Decision and Control 533,
https://doi.org/10.1007/978-3-031-62154-3_4

4.2 Basics of Spatial Grasp Model and Technology

Only main technology ideas are mentioned below, more can be found in the related patent [1], published books [2–10], latest journal papers [11–18], also in numerous previous publications on this paradigm.

Within Spatial Grasp Model and Technology (SGT), a high-level operational scenario expressed in recursive Spatial Grasp Language (SGL), starting in any world point *propagates, covers, and matches the distributed environment in parallel wave-like mode* as symbolically shown in Fig. 4.1, Such propagation can result in returning and analyzing the reached states and data, which may be arbitrarily remote, or used for launching more waves.

The SGL allows for organizing *direct space presence and operations* with unlimited powers and parallelism. Its universal recursive organization, with operational scenarios called *grasp*, can be expressed by a single formula:

$$\boldsymbol{grasp} \;\rightarrow\; \boldsymbol{constant} \mid \boldsymbol{variable} \mid \boldsymbol{rule}(\{\boldsymbol{grasp},\})$$

The *rule* expresses certain action, control, description or context accompanied with operands, which can be any *grasps* too. The rules, starting in certain points, can organize navigation of the world sequentially, in parallel, or any combinations thereof. They can result in the same application points or cause movement to other world points with obtained results left there or returned.

The *SGL interpreter* consists of specialized modules serving SGL scenarios or their parts happened to be inside this interpreter, also organizing exchanges with other interpreters for distributed SGL scenarios. Communicating SGL interpreters can be in arbitrary number of copies, representing *powerful spatial engines operating without central resources or control*, as in Fig. 4.2. They can effectively work with *spatial graph and network data* of any volumes and distributions.

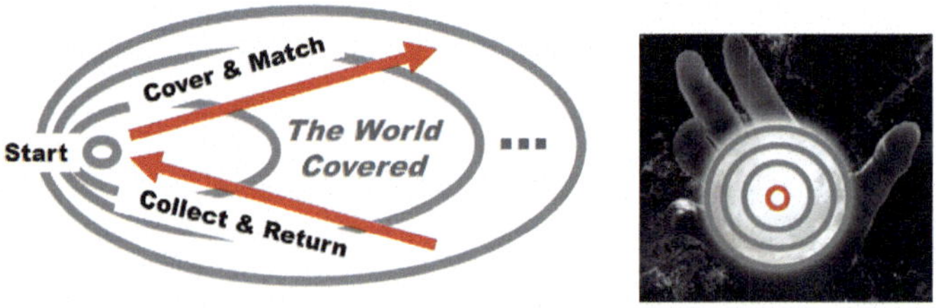

Fig. 4.1 Parallel wavelike world coverage under Spatial Grasp model

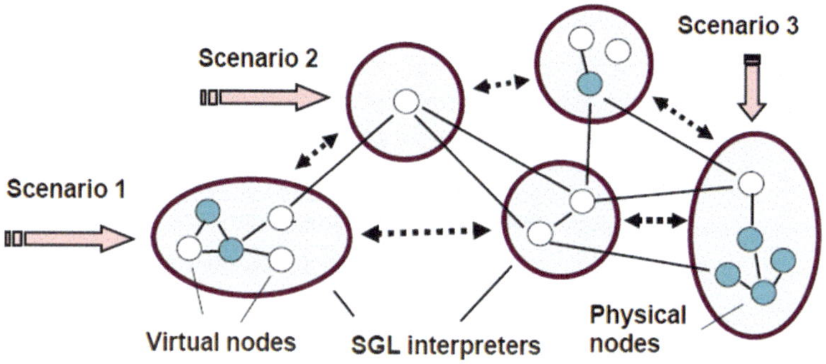

Fig. 4.2 SGL distributed networked interpretation

4.3 Spatial Grasp Language (SGL) Summary

Except mentioned above SGT patent, books and recent papers on SGT and SGL [1–18], there are also many more publications devoted to SGL and its previous versions called WAVE [19–61]. Top language details, its syntax, and main constructs can be summarized as follows.

- **Top SGL summary**

grasp	→	*constant* ǀ *variable* ǀ *rule* ({ *grasp*, })
constant	→	*information* ǀ *matter* ǀ *special* ǀ *custom* ǀ *grasp*
variable	→	*global* ǀ *heritable* ǀ *frontal* ǀ *nodal* ǀ *environmental* ǀ *grasp*
rule	→	*type* ǀ *usage* ǀ *movement* ǀ *creation* ǀ *echoing* ǀ*verification* ǀ *assignment* ǀ *advancement* ǀ *branching* ǀ *transference* ǀ *exchange* ǀ *timing* ǀ *qualifying* ǀ *grasp*

- **SGL syntax and main constructs**

In the language summary below, syntactic categories are shown in italics, vertical bar separates alternatives, parts in braces indicate zero or more repetitions with a delimiter at the right, and constructs in brackets are optional. The remaining characters and words are the language symbols (including boldfaced braces).

grasp	→	*constant* ǀ *variable* ǀ [*rule*] [({ *grasp*,})]
constant	→	*information* ǀ *matter* ǀ *special* ǀ *custom* ǀ *grasp*
information	→	*string* ǀ *scenario* ǀ *number*
string	→	'{*character*}'
scenario	→	{{*character*}}
number	→	[*sign*]{*digit*}[.{*digit*}[e[*sign*]{*digit*}]]

matter	→	"{*character*}"
special	→	`thru` | `done` | `fail` | `fatal` | `infinite` | `nil` | `any` | `all` | `other` | `allother` | `current` | `passed` | `existing` | `neighbors` | `direct` | `forward` | `backward` | `synchronous` | `asynchronous` | `virtual` | `physical` | `executive` | `engaged` | `vacant` | `firstcome` | `unique`
variable	→	*global* | *heritable* | *frontal* | *nodal* | *environmental* | *grasp*
global	→	`G`{*alphameric*}
heritable	→	`H`{*alphameric*}
frontal	→	`F`{*alphameric*}
nodal	→	`N`{*alphameric*}
environmental	→	`TYPE` | `NAME` | `CONTENT` | `ADDRESS` | `QUALITIES` | `WHERE` | `BACK` | `PREVIOUS` | `PREDECESSOR` | `DOER` | `RESOURCES` | `LINK` | `DIRECTION` | `WHEN` | `TIME` | `STATE` | `VALUE` | `IDENTITY` | `IN` | `OUT` | `STATUS` | `SIZE` | `WEIGHT` | `LENGTH`
rule	→	*type* | *usage* | *movement* | *creation* | *echoing* | *verification* | *assignment* | *advancement* | *branching* | *ransference* | *exchange* | *timing* | *qualifying* | *grasp*
type	→	`global` | `heritable` | `frontal` | `nodal` | `environmental` | `matter` | `number` | `string` | `scenario` | `constant`
usage	→	`address` | `coordinate` | `content` | `index` | `time` | `speed` | `name` | `place` | `center` | `range` | `doer` | `node` | `link` | `unit`
movement	→	`hop` | `hopfirst` | `hopforth` | `move` | `shift` | `follow`
creation	→	`create` | `linkup` | `delete` | `unlink`
echoing	→	`state` | `rake` | `order` | `unit` | `unique` | `sum` | `count` | `first` | `last` | `min` | `max` | `random` | `average` | `sortup` | `sortdown` | `reverse` | `element` | `position` | `fromto` | `add` | `subtract` | `multiply` | `divide` | `degree` | `separate` | `unite` | `attach` | `append` | `common` | `withdraw` | `increment` | `decrement` | `access` | `invert` | `apply` | `location` | `distance`
verification	→	`equal` | `nonequal` | `less` | `lessorequal` | `more` | `moreorequal` | `bigger` | `smaller` | `heavier` | `lighter` | `longer` | `shorter` | `empty` | `nonempty` | `belong` | `notbelong` | `intersect` | `notintersect` | `yes` | `no`
assignment	→	`assign` | `assignpeers`
advancement	→	`advance` | `slide` | `repeat` | `align` | `fringe`
branching	→	`branch` | `sequence` | `parallel` | `if` | `or` | `and` | `orsequence` | `orparallel` | `andsequence` | `andparallel` | `choose` | `quickest` | `cycle` | `loop` | `sling` | `whirl` | `split` | `replicate`
transference	→	`run` | `call`

exchange → `input | output | send | receive | emit | get`

timing → `sleep | allowed`

qualification → `contain | release | trackless | free | blind | quit | abort | stay | lift | seize | exit`

The following sections provide detailed explanations of the mentioned above constructs and their practical usage.

4.4 SGL Constants

Constants can be self-identifiable by the way written, as follows, or defined by special rules embracing them with arbitrary textual representations.

- **Information**

A *string* can be represented as a sequence of characters embraced by opening-closing single quotation marks:

```
'{character}'
```

This sequence can contain other single quotes only if appear in opening-closing pairs, with such nesting allowed to any depth. If single words representing information not intersect with other language constructs, the quotes can be omitted. Another string representation may be *explicit SGL scenario body*, where the sequence of characters should be placed into opening-closing braces, shown below in bold to distinguish from braces for textual repetition in the language syntax:

```
{{character}}
```

Braces will indicate the text as a potential *scenario* code which should be optimized *before* its usage.

Number can be represented in a standard way, similar to traditional programming languages, generally in the following form (with brackets identifying optional parts and braces the repeating characters).

```
[sign]{digit}[.{digit}[E[sign]{digit}]]
```

- **Physical matter**

Physical *matter* (physical objects including) in the most general way can be reflected in SGL by a sequence of characters embraced by opening-closing double quotation marks:

```
"{character}"
```

- **Special constants**

These include the following.

Thru—indicates (or artificially sets up) control state of the scenario in the current world point as an *absolute success* with possibility of further scenario evolution from this particular point.

done—indicates (or sets artificially up) control state as a *successful termination* in the current world point, with blocking further scenario development from this particular point.

fail—indicates (or artificially sets up) scenario control state as *failure* in the current world point, without possibility of further scenario development from this particular point.

fatal—indicates (or sets up artificially) control state as nonlocal *fatal failure* starting in the current world point and causing massive termination and removal of all active distributed processes with related data in this and other world points reached by the same scenario. The destructive influence of this state may be contained at higher levels by special rules.

infinite—indicates infinitely large value.

nil—indicates no value at all.

any, **all**, **other**, **allother**—stating that *any one*, *all* (the current one including), *any other*, or *all other* elements can be considered and used by some rule.

Current—refers to the current element (like node) only, for its further consideration or reentering.

passed—informing that the mentioned elements (like world nodes) have already been passed by the current branch to the current point. They may be accessed easier by backwarding via the history-based distributed control than by a global search.

existing—hinting that world nodes with the given names already exist and should not be created again.

Neighbors—stating that the nodes to be accessed are among direct neighbors of the current node, i.e. within a single hop via existing links.

Direct—stating that the mentioned nodes should be accessed or created from the current node directly, without considering semantic links to them (even if such links may already exist).

forward, **backward**, **neutral**—allowing control to move from the current node via existing links along, against, or regardless their orientations (non-oriented links can always be traversed in both directions).

synchronous, **asynchronous**—a modifier setting synchronous or asynchronous mode of operations.

virtual, **physical**, **executive**—indicating or setting the type of a node the scenario is currently dealing with (the node can also be of a combined type).

engaged, **vacant**—indicating or setting the state of a resource the current scenario is dealing with.

`Firstcome`—allows the current scenario with its unique identity to enter the world nodes only first time (the capability based on internal node marking mechanisms).

`Unique`—allows the return of only unique elements received from the embraced scenario final positions while omitting duplicates.

- **Custom constants**

Other self-identifiable, or *custom*, constants can be incorporated to be directly processed by updated SGL, if do not conflict with the language syntax, otherwise should be declared by special rules.

- **Grasp**

Constants can also be represented and produced by individual scenarios of any complexity and space coverage, potentially resulting in data structures of arbitrary complexity, i.e. generally considered as ***grasps*** too.

4.5 SGL Variables

Their types can be defined by how their names are written, but in general, variables may have any names if their types are declared by special rules, as mentioned in Sect. 4.5.

- **Global variables**

This is the most expensive type of SGL variables, with their names starting with capital `G` and followed by sequences of alphabetic letters and/or digits:

```
G{alphameric}
```

These variables can exist only in single copies with particular names, being common for both read and write operations to all processes of the same scenario, regardless of their physical or virtual distribution and world points they may cover (see Fig. 4.3).

- **Heritable variables**

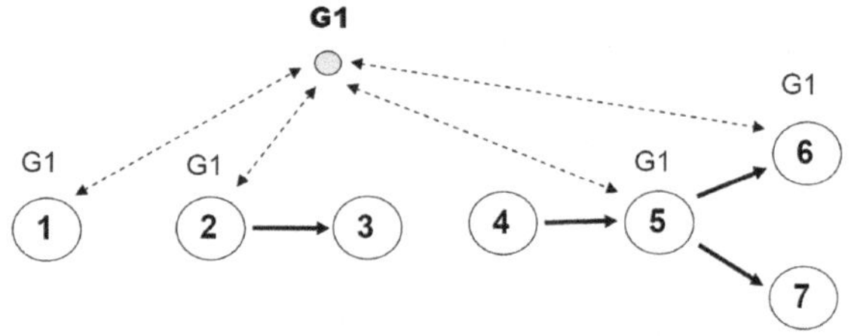

Fig. 4.3 Global variables

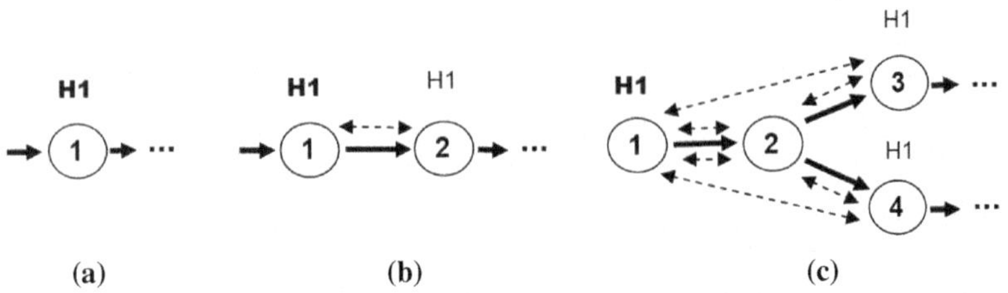

Fig. 4.4 Heritable variables

The names of these variables should start with capital `H` if not defined by a special rule:

```
H{alphameric}
```

Heritable variables, being created by first assignment to them at some scenario development stage, are becoming common for read–write operations *for all subsequent* scenario operations (generally multiple, parallel and distributed) evolving from this particular point and wherever is space they may happen to be. This means that such variables are unique only within concrete hereditary scenario developments, to all their depth (see Fig. 4.4 for sequential stages a, b, c of the scenario evolution).

- **Frontal variables**

These are mobile type variables with names starting with capital `F`, which are propagating in distributed spaces while keeping their contents on the forefronts of evolving scenarios:

```
F{alphameric}
```

Each of these variables is serving only the current scenario branch operating in the current world point. They cannot be shared with other branches evolving in the same or other world points, while always accompanying the scenario control. If the scenario splits into individual branches, these variables are replicated with the same names and contents and serve these branches independently (see Fig. 4.5 for sequential stages a, b, c of the scenario evolution).

- **Nodal Variables**

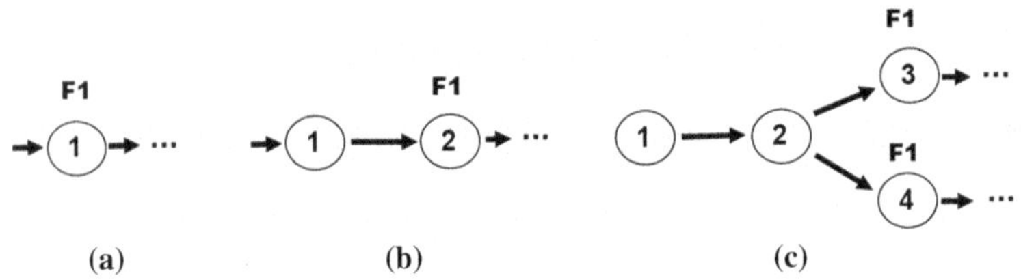

Fig. 4.5 Frontal variables

Variables of this type, their identifiers starting with capital N, are a temporary and exclusive property of the world points visited by SGL scenarios, which can create, change or remove them (see Fig. 4.6):

```
N{alphameric}
```

Capable of being shared by all scenario branches visiting these nodes, they are created by first assignment to them and stay in the node until removed explicitly, or at least a single scenario that passed through this node still remains active. These variables also cease to exist when nodes they associate with are removed by any scenario reaching them.

- **Environmental Variables**

These are special variables with *reserved names* (all in capitals) which allow us to have access to physical, virtual, and execution worlds when they are navigated by SGL scenarios, also to some parameters of the language interpretation system itself.

TYPE—indicates the type of a node the current scenario step associates with and returns a verbal expression of the node's type (i.e. `virtual`, `physical`, `executive` or their combination). It can also change the existing node's type by assigning to it another value (simple or combined).

NAME—returns name of the current node as a string of characters (only if the node has virtual or executive dimension or both). Assigning to this variable when staying in the node can change the node's name.

CONTENT—returns content of the current node (if it has virtual or executive dimension or both) as any text, vector, or nested structure of multiple texts, etc. Assigning to this variable when staying in the node can set up or change the node's content.

ADDRESS—returns a unique address of the node having virtual dimension. This is read-only variable as node addresses are set up automatically by the underlying distributed SGL interpretation system during node's creation, or by an external system (like internet address of the node).

QUALITIES—identifies a list of selected formalized physical parameters associated with the current physical position, or node, depending on the chosen implementation and application (for example, these may be temperature, humidity, air pressure, visibility, radiation, noise or pollution level, density, salinity, etc.). These

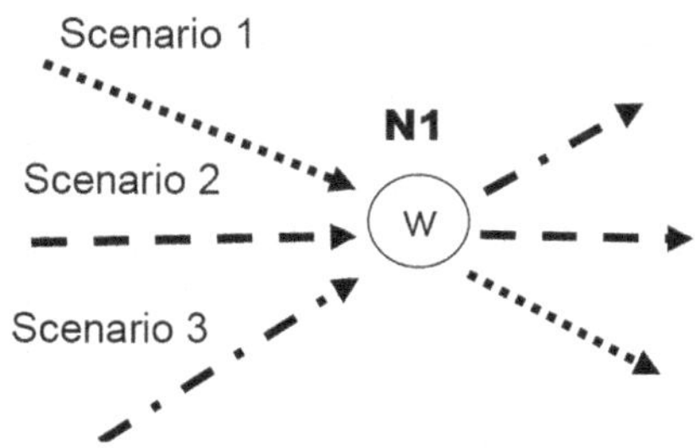

Fig. 4.6 Nodal variables

parameters (generally as a list of values) can be obtained by reading the variable, and some can be changed, thus influencing the world from its particular point.

WHERE—keeps world coordinates of the current physical node (or the one having physical dimension too) in the chosen coordinate system. These coordinates can be obtained by reading this variable. Assigning a new value to this variable (with possible speed added) can cause physical movement into the new position.

BACK—keeps internal system link to the preceding world node (virtual, executive or combined one) allowing the scenario to most efficiently return to the previously occupied node, if needed.

PREVIOUS—refers to the absolute and unique address of the previous virtual node (or combined one with executive and/or physical dimensions), allowing us to return to the node directly. Its content, unlike with `BACK`, can be lifted, recorded, and used elsewhere in the scenario (but not changed).

PREDECESSOR—refers to the name of preceding world node (the one with virtual or executive dimension, visited just before the current one). Its content can be lifted, recorded and subsequently used, including for organization of direct hops to this node (on most expensive level, however).

DOER—keeps the name of the device (say, laptop, robot, satellite, smart sensor, or a specially equipped human) which interprets the current SGL code in the current world position. This device can be initially chosen for the scenario automatically from the list of recommended devices or just picked up from those expected available. It can also be appointed explicitly by assigning its name to `DOER`.

RESOURCES—may keep a list of available or recommended resources (human, robotic, electronic, mechanical etc., by their types or names) which can be used for planning and execution of the current and subsequent parts of the SGL scenario. This list can also contain potential doers which may appear (by their names) in variables `DOER`.

LINK—keeps the name (same as content) of the virtual link which has just been passed. Assigning a new value to it can change the link's content/name. Assigning `nil` or empty to `LINK` removes the link passed.

DIRECTION—keeps direction (along, against, or neutral) of the passed virtual link. Assigning to this variable values like `plus`, `minus`, or `nil` (same as +, -, or empty) can change its orientation or make the link non-oriented.

WHEN—assigning value to this variable sets up an absolute starting time for the following scenario branch (i.e. starting with the next operation), thus allowing us to suspend and schedule operations and their groups in time.

TIME—returns the current absolute system time as the read-only global variable.

STATE—can be used for explicit setting resultant control state of the current scenario step by assigning to it one of the following constants: `thru`, `done`, `fail`, or `fatal`, which will influence further scenario development from the current world point (and in a broader scale in the case of `fatal`). These control states are also generated implicitly and automatically on the results of success or failure of different operations (belonging to the internal interpretation mechanisms of SGL scenarios). Reading `STATE`will always return `thru` as this could be possible only if the previous operation terminated with `thru` too, thus letting this operation to proceed.

VALUE—when accessed, returns the resultant value of the latest, i.e. preceding, operation (say, an assignment to it or any other variable, unassigned result of arithmetic or string operation, or just naming a variable or constant).

IDENTITY—keeps identity, or color, of the current SGL scenario or its branch, which propagates together with the scenario and influences grouping of different nodal variables under this identity at world nodes reached. This allows different scenarios or their branches with personal identities to be protected from influencing each other, even if they are using same named nodal variables in the same world nodes.

IN—special variable requesting and reading data from the outside world in its current point. The received data is becoming the resultant value of the reading operation.

OUT—special variable allowing us to issue information from the scenario in its current point to the outside world, by assigning the output value to this variable.

STATUS—retrieving or setting the status of (especially doer) node in which the scenario is currently staying (like `engaged` or `vacant`, possibly, with a numerical estimate of the level of engagement or vacancy). This feedback from implementation layer on the SGL scenario layer can be useful for a higher-level supervision, planning and distribution of resources executing the scenario rather than doing this implicitly and automatically.

SIZE, **WEIGHT**, **LENGTH**—correspondingly, size of the node and weight and length of the link, as additional parameters representing certain real world components.

- **Grasp**

Variables can also be represented and declared by individual scenarios of any complexity and space coverage, potentially resulting in names or collection of names of variables and their contents be processed together, i.e. generally considered as ***grasps*** too.

4.6 SGL Rules

- **Type**

These rules explicitly assign types to different constructs, with their existing repertoire following.

global, **heritable**, **frontal**, **nodal**, **environmental**—allow different types of variables to have any alphanumeric names rather than those oriented on self-identification, as explained before. These names will represent variables with needed types in the subsequent scenario developments unless redefined by these rules too.

matter, **number**, **string**, **scenario**, **constant**—allow arbitrary results obtained by the embraced scenario to properly represent the needed values rather than using self-identifiable representations mentioned before.

- **Usage**

These rules explain how to use the information units they embrace.

address—identifies the embraced value (which may also be arbitrary scenario producing this value) as an address of a virtual node.

coordinate—identifies the embraced value as physical coordinates (say, one, two or three dimensional).

content—identifies the embraced operand as a content (or contents).

index—identifies the embraced operand as an index (or indices) which may represent orders of elements in a list for its search by the index operation.

time—informs that the embraced operand represents time value.

speed—informs that the embraced operand represents a value of speed.

name—identifies the embraced operand as a name (say, of a virtual or executive node or nodes).

center—depending on applications, indicates that virtual address or physical coordinates embraced may relate to the center of some region.

range—identifies virtual or physical distance that can, for example, be used as a threshold for certain operations in distributed spaces, especially those evolving from a chosen or expected center.

doer—identifies the embraced name or any other value as belonging to executive node (like human, robot, server, satellite, smart phone, etc.).

node—(or **nodes**, if more appropriate) identifies the embraced value or values as keeping names of nodes having virtual or/and executive dimensions.

link—(or **links**, when more appropriate) informs that the embraced value or values represent names of links connecting nodes.

- **Movement**

It may result in virtual hopping to the existing nodes (the ones having virtual or/and executive dimensions) or in real movement to new physical locations, subsequently starting the remaining scenario if any (with current frontal variables and control) in the nodes reached. Some of the movement examples are shown in Fig. 4.7.

hop—sets electronic propagation to a node (or nodes) in virtual, execution, or combined spaces (the latter may have physical dimension too), directly or via semantic links connecting them with the starting node. In case of a direct hop, except

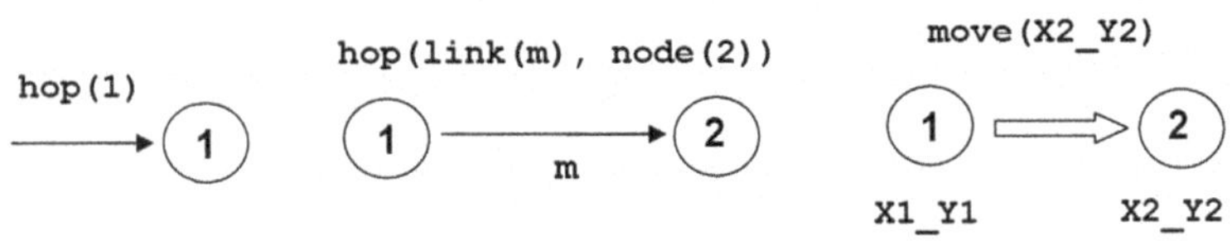

Fig. 4.7 Examples of movement in SGL

destination node name or address, special modifier `direct` may be included into parameters of the rule. If the hop is to take place from a node to a particular node via existing link, both destination node name/address and link name (with orientation if appropriate) should be parameters of the rule.

hopfirst—modification of the `hop` rule allowing it to come to a node only first time (for the scenario with certain identity), which is based on internal interpretation mechanism marking the nodes visited.

hopforth—modification of the previous rule allowing it to hop to a node which is not the one just visited before, i.e. excluding the return to the previous node. It may be considered as a restricted variant of `hopfirst` rule.

move—sets real movement in physical world from the current node with physical dimension (which may be combined with virtual and executive ones) to a particular location given by coordinates in a chosen coordinate system. Speed value for the physical propagation by `move` may be given as an additional parameter.

shift—differs from the `move` only in that movement in physical world is set by deviations of physical coordinates from the current position rather than by their absolute values.

follow—allows us to move in virtual, physical and combined spaces using already recorded and saved paths from a starting node to the destinations reached, to enter the latter again.

- **Creation**

These rules create or remove nodes and/or links leading to them during distributed world navigation. After termination of the rules their resultant values will correspond to the names of reached nodes with states `thru` in them, and the next scenario steps, if any, may start from all these nodes. If the creation or removal operation fails, its resultant value will be `nil` and control state `fail` in the node the rule started. Related creative examples are shown in Fig. 4.8.

create—starting in the current world position, creates either new isolated nodes, or new virtual link-node pairs. For the first case, the rule may additionally use modifier `direct` explicitly indicating direct nodes creation. For the second case, the rule is supplied with names and orientations of new links and names of new nodes these links should lead to, which may be multiple.

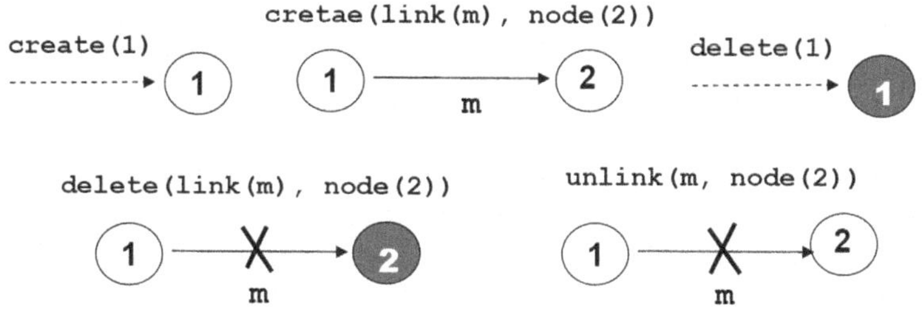

Fig. 4.8 Creation and removal of nodes and/or links

Fig. 4.9 Scenario echo rules

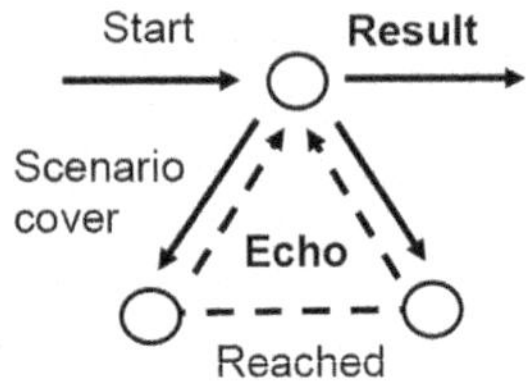

linkup—restricts the previous rule by creating only links with proper names from the current node to the already existing nodes given by their names or addresses.

delete—starting from the current node, removes links together with nodes they should lead to. Links and nodes to be removed should be either explicitly named or represented by modifiers `any` or `all`. Using modifier `direct` will allow us to remove such node (or nodes) from the current node directly.

unlink—removes only links leading to neighboring nodes where, similar to the previous case, they should be explicitly named or modifiers `any` or `all` used instead.

- **Echoing**

This class of rules, oriented on various aspects of data and knowledge processing, contains the following rules which may use local and remote values for different operations. The listed rules use terminal world positions reached by the embraced scenario with their control states and associated final values (which may be local or arbitrarily remote) to obtain the resultant state and value in the location where the rule started (as in Fig. 4.9).

state—returns the resultant generalized state of the embraced SGL scenario upon its completion, whatever its complexity and space coverage may be. This state being the result of ascending fringe-to-root generalization of terminal states of the scenario embraced, where states with higher power (their sequence from maximum to minimum values as: `fatal`, `thru`, `done`, `fail`) dominate in this potentially distributed and parallel process.

rake—returns a list of final values of the scenario embraced in arbitrary order, which may, for example, be influenced by the order of completion of branches and times of reaching their final destinations. Additionally using `unique` as modifier, described before, the rule will result in collecting only unique values.

order—returns an ordered list of final values of the scenario embraced corresponding to the order of launching related branches rather than the order of their completion. For potentially parallel branches, these orders may, for example, relate to how they were activated, possibly, with the use of time stamps upon invocation.

unit—returns a list of values while arranging it as an integral parenthesized unit which should not be mixed with elements returned from other branches which may represent integral units too, to form (potentially hierarchical and nested) lists of lists of the obtained values at higher levels.

sum—returns the sum of all final values of the scenario embraced.

count—returns the number of all resultant values associated with the scenario embraced.

first, **last**, **min**, **max**, **random**, **average**—return, correspondingly, the first, last, minimum, maximum, randomly chosen, or average value from all terminal values returned by the scenario embraced.

sortup, **sortdown**—return an ordered list of values produced by the embraced scenario operand, starting from minimum or maximum value and ending, correspondingly, with maximum or minimum one.

reverse—changes to the opposite the order of values from the embraced operand.

element—returns the value of an element of the list on its left operand requested by index or content given by the right operand. If the right operand is itself a list of indices or contents, the result will be the list of corresponding values from the left operand. If `element` is used within the left operand of assignment, instead of returning values it will be providing an access to them in order to be updated.

position—returns the index (or indices) of the list on its left operand requested by the content given by the right operand. There may be more than a single index returned if same content repeats in the list, or if the right operand is itself a list of contents.

fromto—returns an ordered list of digital values by naming its first (operand 1) and last (operand 2) elements as well as step value (operand 3) allowing the next element to be obtained from the previous one. Another modification (depending on implementation) may take into account the starting element, step value, and the number of needed elements in the list.

add, **subtract**, **multiply**, **divide**, **degree**—perform corresponding operations on two or more operands embraced, each potentially represented by arbitrary scenario with local or remote results. If the operands themselves provide multiple values, as lists, these operations are performed between peer elements of these lists, with the resultant value as a list too.

separate—separates the left operand string value by the string at the right operand used as a delimiter in a repeated manner for the left string, with the result being the list of separated substring values. If the right operand is a list of delimiters, its elements will be used sequentially and cyclically unless the string at the left is fully processed/partitioned.

unite—integrates the list of values at the left (as strings, or to be converted into strings automatically) by a repeated delimiter as a string too (or a cyclically used list of them) at the right into a united string.

attach—produces the resultant string by connecting the right string operand directly to the end of the left one. If operands are lists with more than one element, the attachment is made between their peer elements, receiving the resultant list of united strings. This rule can also operate with more than two operands.

append—forms the resultant list from left and right operands by appending the latter to the end of the former as individual elements, where both operands may be lists themselves. More than two operands can be used too for this operation.

common—returns intersection of two or more lists as operands, with the result including only same elements of all lists, if any.

withdraw—its returned result will be the first element of the list provided by the embraced operand, along with withdrawing this element from the head of the list (thus simultaneously changing the content of the variable). This rule can have another operand providing the number of elements to be withdrawn in one step and represented as the result.

increment—adds 1 (one) to the value of the embraced operand which will be the result on this rule, thus simultaneously changing the content of the operand itself. If another value, not 1, to be added, the second operand can keep this value.

decrement—behaves similar to the previous rule `increment` but subtracts rather than adds 1 from the value of the embraced operand, with the content of the latter simultaneously changed. Second operand can be used if the value to subtract not equals 1.

access—by embracing a scenario or its branch, returns a reference to the internal history-based optimized and recorded structure leading from the rule-activation node to the reached terminal nodes. This reference can be remembered (say, in a variable) and subsequently used from the same starting node for reaching exactly the same terminal nodes again.

invert—changes the sign of a value or orientation of a link to the opposite, while producing no effect on zero values or non-oriented links.

apply—organizes application of the first operand as one or a set of rules described above operating jointly from the same starting point to the same second scenario operand, which may be arbitrary. If multiple application rules engaged on the first operand, the obtained results on the second operand can be multiple too.

location—returns world locations of the final nodes reached by the embraced scenario, which mean for virtual nodes their network addresses, and for physical nodes physical coordinates.

distance—returns distance between two physical points defined by absolute physical coordinates expressed by its parameters, where each one can be represented by an arbitrary scenario.

- **Verification**

These rules provide control state `thru` or `fail` reflecting the result of concrete verification procedure, also `nil` as own resultant value, while remaining after completion in the same world positions where they started, as in Fig. 4.10.

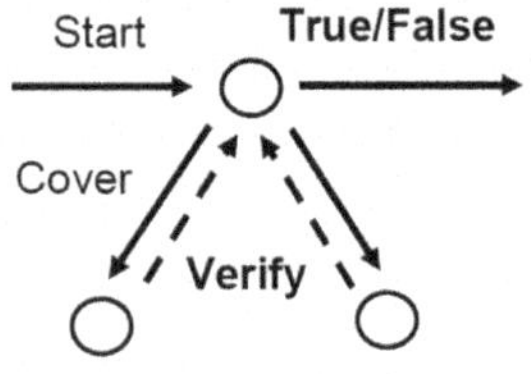

Fig. 4.10 Verification rules

equal, **nonequal**, **less**, **lessorequal**, **more**, **moreorequal**, **bigger**, **smaller**, **heavier**, **lighter**, **longer**, **shorter**—make corresponding comparison between left and right operands, which can represent information or physical matter/objects, or both. In case of vector operands, state `thru` appears only if all peer values satisfy the condition set up by the rule (except `nonequal`, for which even a single non-correspondence between any peers will result in overall `thru`).

empty, **nonempty**—checks for emptiness (i.e. non-existence of anything, same as `nil`) or non-emptiness (existence) of the resultant value obtained from the embraced scenario.

belong, **notbelong**—verifies whether the left operand value (single or a list with all its elements) belongs as a whole to the right operand generally represented as a list (which may have a single element too).

intersect, **notintersect**—verifies whether there are common elements (values) in the left and right operands, considered generally as lists. More than two operands can be used for these rules too, with at least a single same element to be present in all of them to result `thru` for `intersect`, or no such elements for `notintersect`.

yes—verifies generalized state of the embraced scenario providing own control state `thru` in case of `thru` or `done` from the entire scenario, and control state `fail` in case of resultant `fail` or `fatal` (thus allowing to continue from the node where the rule started only in case of success of the embraced scenario.

no—verifies generalized state of the embraced scenario resulting with own control state `thru` in case of `fail` or `fatal` from the scenario, and control state `fail` in case of `thru` or `done` (i.e. allowing to continue from the rule's starting node only in case of failure of the embraced scenario).

- **Assignment**

These rules assign the result of the right scenario operand (which may be arbitrarily remote, also represent a list of values which can be nested) to the variable or set of variables directly named or reached by the left scenario operand, which may be remote too (see Fig. 4.11). The left operand can also provide pointers to certain elements of the reached variables which should be changed by the assignment rather than the whole contents of variables.

assign—assigns the same value of the right operand (which may be a list of values) to all values (like, say, node names) or variables accessed by the left operand (or their particular elements pointed, which may themselves become lists

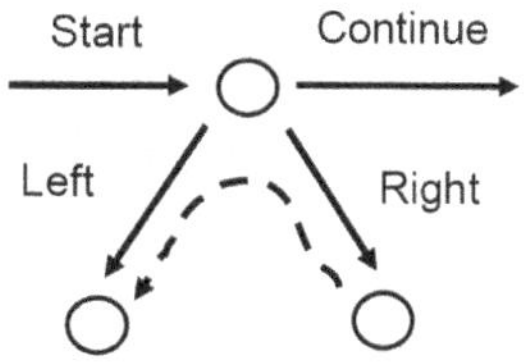

Fig. 4.11 Assignment rules

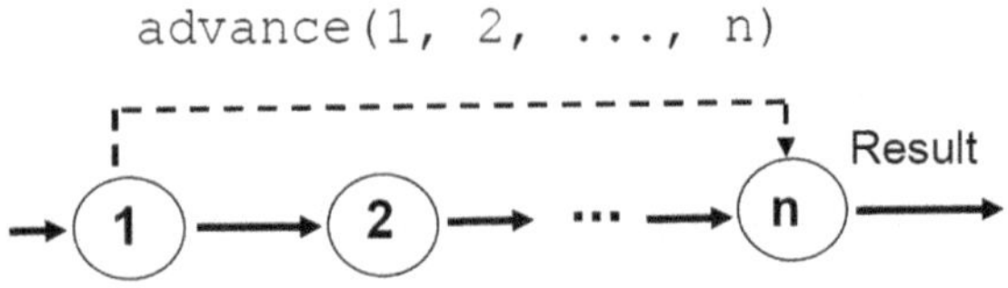

Fig. 4.12 Advancement rules

after assignment, thus extending the lists of contents of these variables). If the right operand is `nil` or empty, the left operand nodes or variables as a whole (or only their certain elements pointed) will be removed.

`assignpeers`—assigns values of different elements of the list on the right operand to different values or variables (or their pointed elements) associated with the destinations reached on the left operand, in a peer-to-peer mode.

- **Advancement**

These rules can organize forward or "in depth" advancement in space and time of the embraced scenarios separated by comma, as in Fig. 4.12. They can evolve within their sequence in synchronous or asynchronous mode using modifiers `synchronous` or `asynchronous` (the second one optional, as asynchronous being a default mode).

`advance`—organizes stepwise scenarios advancement in physical, virtual, executive or combined spaces, also in a pure computational space (the latter when staying in the same world nodes with certain data processing, thus moving in time only). For this, the embraced SGL scenario-operands are used in a sequence, as written, where each new scenario shifts to and applies from all terminal world points reached by the previous scenario. The resultant world positions and values on the whole rule are associated with the final steps of the last scenario on the rule. And the rule's resultant state is a generalization of control states associated with these final steps.

`slide`—works similar to the previous rule unless a scenario in their sequence fails to produce resultant state `thru` or `done` from some world node. In this case the next scenario from the sequence will be applied from the same starting position of the previous, failed, scenario, and so on. The resultant world nodes and values in them will be from the last successfully applied scenarios (not necessarily the same from their sequence, as independently developing in different directions).

`repeat`—invokes the embraced scenario as many times as possible, with each new iterations taking place in parallel from all final positions with state `thru` reached by the previous invocations (see Fig. 4.13). If some scenario iteration fails, its current starting position with its value will be included into the set of final positions and values on the whole rule (this set may have starting positions from different failed iterations which developed independently in a distributed space). By supplying additional numeric modifier to this rule, it is possible to explicitly limit the number of allowed scenario repetitions.

`align`—is based on confirmation of full termination of all activities of the embraced operand-scenario in all its final nodes. Only after this, the remaining scenario part, if any, will be allowed to continue from all the nodes reached.

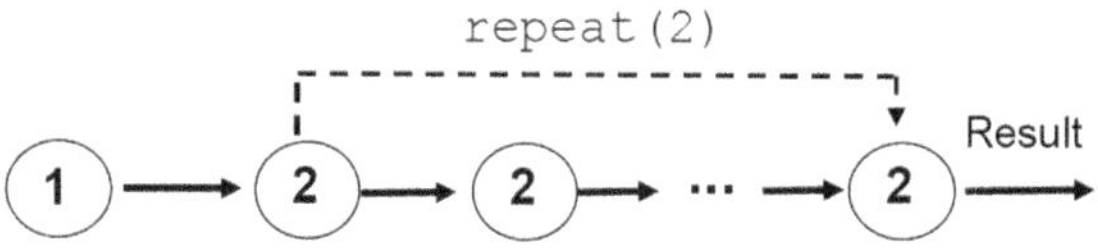

Fig. 4.13 Scenario repetition

fringe—allows us to establish certain constraints (say, by additional parameters) on the terminal world nodes reached by the embraced scenario with final values in them, to be considered as starting positions for the following scenario parts. For example, by comparing values in all terminal nodes and allowing the scenario to continue from a node with maximum or minimum value, integrating this rule with previously mentioned rules `max` or `min` like **max_fringe** or **min_fringe** can also be possible.

For the advancement rules, frontal variables propagate on the forefronts together with advancement of control and operations in distributed spaces, with next scenarios or their iterations picking up frontal variables brought to their starting points by the previous scenarios, being also replicated if this control automatically splits into different branches.

- **Branching**

These rules allow the embraced set of scenario operands to develop "in breadth", each from the same starting position (as in Fig. 4.14), with the resultant set of positions and order of their appearance depending on the logic of a concrete branching rule The rest of the SGL scenario will be developing from all or some of the positions and nodes reached on the rule. The branching may be static and explicit if we have a clear set of individual operand scenarios separated by comma. It can also be implicit and dynamic, as explained later. For all branching rules that follow, the frontal variables associated with the rule's starting position will be replicated together with their contents and used independently within different branches.

branch—the most general and neutral variant of branching, with logical independence of the scenario operands from each other and any possible order of their invocation and development from the starting position. The resultant set of positions reached with their associated values will unite all terminal positions & values on all

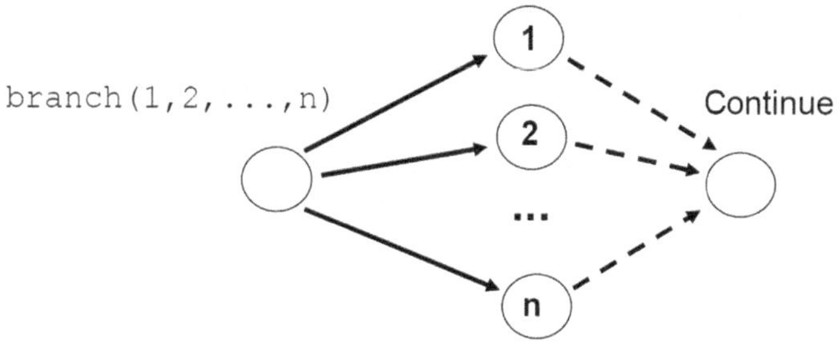

Fig. 4.14 The use of branching rule

scenario operands involved under `branch`. The resultant control state on the whole rule will be based on generalization of the generalized control states on all scenario branches.

`sequence`—organizing strictly sequential invocation of all scenario operands regardless of their success or failure, and launching the next scenario only after full completion of the previous one. The resultant set of positions, values, and rule's global control state will be similar to `branch`. However, the final results may vary due to different invocation order of the scenario operands and possible common information used.

`parallel`—organizing fully parallel development of all scenario operands from the same starting position. The resultant set of positions, values, and rule's control state will be similar to the previous two rules, but may not be the same, as explained before.

`if`—may have three scenario operands. If the *first* scenario results with generalized termination state `thru` or `done`, the *second* scenario is activated, otherwise the *third* one will be launched. The resultant set of positions & associated values will be the same as achieved by the second or third scenarios after their completion. If the third operand-scenario is absent and the first one results with `fail`, or only the first operand is present regardless of its success or failure, the resultant position will be the one the rule started from, with state `thru` and value it had at the start.

`or`—allows *only one* operand scenario with the resulting state `thru` or `done`, without any predetermined order of their invocation, to be registered as resultant, with the final positions & associated values on it to be the resulting ones on the whole rule. The activities of all other scenario operands and all results produced by them will be terminated and cancelled.

`orsequence`—will launch the scenario operands in strictly sequential manner, one after the other as they are written, waiting for their full completion before activating the next operand, unless the first one in the sequence replies with generalized state `thru` or `done` (providing the result on the rule as a whole). Invocation of the remaining scenarios in the sequence will be skipped.

`orparallel`—activates all scenario operands in parallel from the same current position, with the first one in time replying with generalized `thru` or `done` being registered as the resultant branch for the rule. All other branches will be forcefully terminated without waiting for their completion (or just ignored).

`and`—activates all scenario operands from the same position, without any predetermined order, demanding all of them to return generalized states `thru` or `done`. If at least a single operand returns generalized `fail`, the whole rule results with state `fail` and `nil` value in the starting position while terminating the development of all other branches which may still be in progress. If all operand scenarios succeed, the resulting set of positions unites all resultant positions on all scenario-operands with their associated values.

`andsequence`—activates scenario-operands from the same position in the written order, launching next scenario only after the previous one completes with `thru` or `done`, and terminating the whole rule when the current scenario results with fail. The remaining scenario operands will be ignored, and all results produced by this and all previous operands will be removed.

andparallel—activates in parallel all scenario operands from the same world position, terminating the rule when the first one in time results with fail, while aborting activity of all other operands and removing all results produced by the rule.

choose—chooses a scenario branch in their sequence *before* its execution, using additional parameters among which, for example, may be its numerical order in the sequence (or a list of such orders to select more than one branch). This rule can also be aggregated with other rules like first, last or random, by forming combined ones: **choose_first**, **choose_last**, **choose_random**.

quickest—selects the first branch in time replying its complete termination, regardless of its generalized termination state, which may happen to be fail too, even though other branches (to be forcefully terminated now) could respond later with thru or done. The state, set of positions on this selected branch, and their associated values (if any) will be taken as those for the whole rule. This rule assumes that different branches are launched independently and in parallel.

cycle—repeatedly invokes the embraced scenario from the same starting position (as in Fig. 4.15) until its resultant generalized state remains thru or done, where on different invocations the same or different sets of resultant positions (with same or different values) may emerge. The resultant set of positions on the whole rule will be an integration of all positions on all successful scenario invocations with their associated values. The next scenario will be developing from all these world positions reached.

loop—differs from the previous rule in that the resultant set of positions on it being only the set produced by the *last* successful invocation of the embraced scenario.

sling—invokes repeatedly the embraced scenario until it provides state thru or done, always resulting in the same starting position with state thru and its previously associated value.

whirl—endlessly repeating the embraced scenario from the starting position regardless of its success or failure and ignoring any resultant positions or values produced. External forceful termination of this construct may be needed, like using first in time termination of another, competitive, branch.

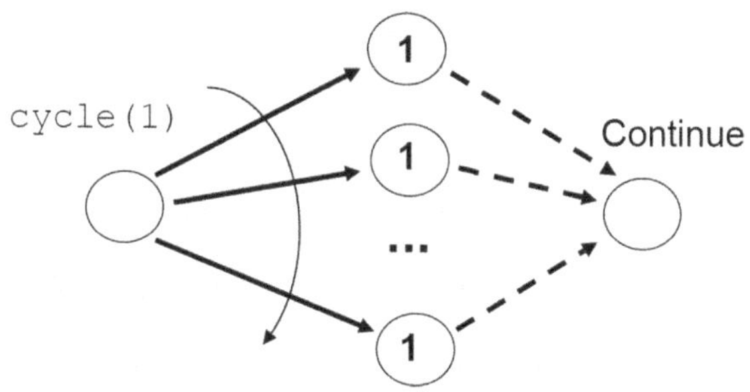

Fig. 4.15 The use of cycling rule

split—performs, if needed, additional (and deeper than usual) static or dynamic partitioning of the embraced scenario into different branches, especially in complex and not clear at first sight cases, all starting from the same current position. It may be used alone or in combination with the above mentioned branching rules while preparing separate branches for these rules, ahead of their invocation.

replicate—replicates the scenario given by its second operand, providing the number of its copies given by the first operand, with each copy behaving as independent branch starting from the same current world position.

- **Transference**

The rules organize transference of control in distributed scenarios.

run—transfers control to the SGL code treated as a procedure and being a result of invocation of the embraced scenario. The procedure (or a list of them) obtained and activated in such a way can produce a set of world positions with associated values and control states as the result on the rule, similar to other rules.

call—transfers control to the code produced by the embraced scenario which may represent activation of external systems. The resultant world position on `call` will be the same where the rule started, with value in it of what has been returned from the external call and state `thru` if the call was successful.

- **Exchange**

These rules include the following.

input—provides input of external information or physical matter (objects) on the initiative of SGL scenario, resulting in the same position but with value received from the outside. The rule may have an additional argument clarifying a particular external source from which the input should take place.

output—outputs the resultant value obtained by the embraced scenario, which can be multiple, with the same resultant position as before and associated value sent outside (in case of physical matter, the resultant value may depend on the applications). The rule may have an additional pointer to a particular external sink.

send—staying in the current position associated with physical, virtual, executive (or combined) node, sends information or matter obtained by the scenario on the first operand to other similar node given by name, address or coordinates provided by the second operand, assuming that a companion rule `receive` is already engaged there. The rule may have an additional parameter setting acceptable time delay for the consumption of this data at the receiving end.

receive—a companion to rule `send`, naming the source of data to be received from (defined similarly to the destination node in `send`). Additional timing (as a second operand) may be set too, after expiration of which the rule will be considered as failed. In case of successful receipt of the data, the rule will result in the same world position and the value obtained.

emit—depending on implementation and technical capabilities, can trigger nonlocal to global continuous broadcasting of the data obtained by the embraced scenario, possibly, with tagging of this source, also setting the emission frequency).

Another operand providing time allowed for this broadcasting may be present too. No feedback from possible consumers of the sent data is expected.

get—tries to receive data which can be broadcast from some source (say, identified by its tag or frequency), with resultant value as the received data and state `thru`. No synchronization with the data emitting node is expected.

- **Timing**

These rules are dealing with conditions related to a time interval for the scenarios they embrace.

sleep—establishes time delay defined by the embraced scenario operand, with suspending activities of this particular scenario branch in the current node. The rule's starting position and its existing value, also state `thru`, will be the result on the rule after the time passed.

allowed—sets time limit by the first operand for an activity of the scenario on the second operand. If the scenario terminates before expiration of this time frame, its resultant positions with values and states will define the result on this rule. Otherwise the scenario will be forcibly aborted.

- **Qualification**

These rules are providing certain qualities or abilities, also setting constraints or restrictions to the scenarios they embrace, as follows.

contain—restricts the spread of abortive consequences caused by control state `fatal` within the ruled scenario. This state may appear automatically and accidentally in different scenario development points or can be assigned explicitly to environmental variable `STATE`, triggering emergent completion of multiple scenario processes and removal of temporary data associated with them. The resultant position on the rule `contain` having state `fatal` inside its scenario will be the one it started from, with value `nil` and state `fail`. If `fatal` did not appear inside this rule, its result will be defined by the completion of the embraced scenario.

release—allows the embraced scenario develop free from the main scenario, abandoning bilateral control links with it (the main scenario after the rule's activation "will not see" this construct any more). The released, now independent, scenario will develop as a whole in the usual way, using its standard subordination and control mechanisms.

trackless—allows the whole embraced scenario to develop absolutely free from any previous stages (i.e. without saving any control and information links with them) like a real virus, being an extreme and global variant of `release`. Under this rule, frontal variables carrying information between different stages will work as usual.

free—differs from the previous case in that despite its independence and control freedom from the main scenario, as before, the embraced scenario will nevertheless be obliged to return the final data obtained in its terminal positions to the main scenario.

blind, quit, abort—after full completion of the embraced scenario, these rules result in the same position the rule started with respective states `done`, `fail`, or `fatal`, thus preventing further scenario development from this point (also triggering nonlocal termination and cancellation processes in case of `fatal`).

`stay`—whatever the scenario embraced and its evolution in space, the resultant position will always be the same this rule started from, with value `nil` and state `thru` in it. If the ruled scenario is omitted, this rule standing alone just represents an empty operation in the current point, which can also be used for declaring a new scenario branch.

`lift`—lifts blocking of the further scenario developments set up by states `done` in the embraced scenario wherever it happened to emerge, substituting them with `thru` and allowing further developments from all such positions.

`seize`—establishes, or seizes, an absolute control over the resources associated with the current virtual, physical, executive or combined node, blocking these from any other accesses and allowing only the embraced scenario to work with them (thus preventing possible competition for the node's assets which may lead to unexpected results). This resource blockage is automatically lifted after the embraced scenario terminates. If more than two scenarios are competing for the node's resources, they will be organized in a FIFO manner at the node.

`exit`—terminates the innermost loop in which it is included.

- **Grasp**

The rules can also be represented by scenarios of any complexity and space coverage that provide names of rules or their collections which should (individually or jointly) manage the embraced scenarios.

4.7 Elementary Examples of Programming in SGL

Only elementary graph and network operations in SGL are shown below, as many of them have been discussed and implemented in the previous SGT and SGL based publications, including [1–39], and will also be analyzed in the following chapters.

Creating isolated virtual node with some name `a` (see Fig, 4.16a, top):

```
create(node(a)) or shortened: create(a)
```

Adding nodes and links. Starting in node `a`, adding link named `r` leading to node `b` to be created too, will be as (see Fig, 4.16a, middle):

```
hop(a); create(link(r), node(b))
```

Creating the two linked nodes from scratch can be done by:

```
advance(create(node(a)), create(link(r), node(b)))
```

or just using traditional semicolon delimiter-operation instead of rule `advance`:

```
create(node(a)); create(link(r), node(b))
```

Maximally shortened, also using `create` for global creative coverage:

```
create(a; (r, b))
```

Providing additional details of nodes and links connecting them can be mentioned as follows (see Fig. 4.16a, bottom):

```
create(node(a, WHERE:X1_Y1, SIZE:big);
    (link(r, DIRECTION:+, WEIGHT:10, LENGTH:50),
     node(b, WHERE:X2_Y2, SIZE:small)))
```

Creating arbitrary network, as in Fig. 4.16b (assuming for simplicity all its links are named `r`) will be as follows.

Step 1: Network topology description naming nodes separated by a colon from the lists of their direct neighbors.

```
Top = (a:(b,c,d,g), b:(a,c,d,e), c:(a,b,f,g),
    d:(a,b,e,g), e:(b,d,f), f:(e,c,g),
    g:(a,c,d,f))
```

Step 2: Separating topology description into individual and parallel control points (using special environmental variable `VALUE`), each registering particular node with its neighbors the latter put into variable `Next` in each point. It then creates this node from this point, also waits for full completion of this parallel step.

```
align(split(Top); Next = VALUE:2; create(VALUE:1));
```

Step 3: Staying in each newly created node, it tries to connect it with links `r` to direct neighbors registered in `NEXT`. As all this operations are performed in parallel in all nodes, to protect from creation of the same links twice, priority is given to the neighbor having its name stronger, using for this environmental variables `PREDECESSOR`, `NAME`, and `BACK` (could be in opposite too).

```
hop(Next); NAME > PREDECESSOR; linkup(r, BACK)
```

The full solution will look like:

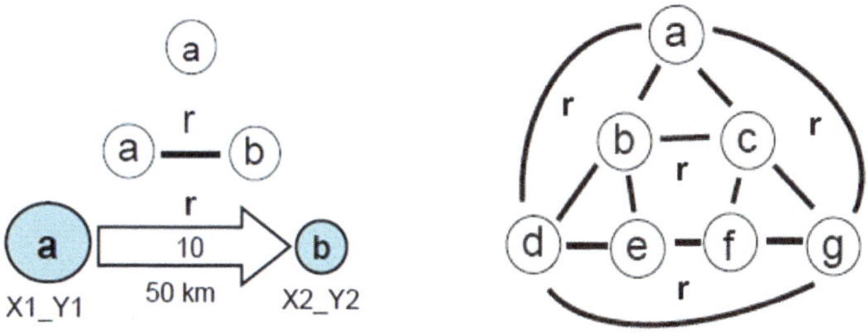

(a) **(b)**

Fig. 4.16 Elementary network operations

```
nodal(Top) = (a:(b,c,d,g), b:(a,c,d,e), c:(a,b,f,g),
        d:(a,b,e,g), e:(b,d,f), f:(e,c,g),
        g:(a,c,d,f));
align(split(Top); frontal(Next) = VALUE:2;
   create(VALUE:1));
hop(Next); NAME > PREDECESSOR; linkup(r, BACK)
```

Reaching nodes from nodes. Having created a network infrastructure like the one as in Fig. 4.16,b, we may, starting in some node like g, reach any other node like b, by the following SGL scenario navigating the network in parallel wavelike mode, without cycling.

```
hopfirst(g);
repeat(hopfirst(all_links); if(NAME == b, done))
```

The rule hopfirst enters each node only once, with marking them, thus blocking possible cycles during the network navigation, also not guaranteeing all possible paths between given nodes to be processed.

Finding all paths between nodes. It is also very easy to find any paths in networks using SGL between nodes of interest. For example, all simple paths from node g to node b (i.e. with non repeating nodes) can be easily found from Fig. 4.16b and finally printed by:

```
frontal(Path); hop(g);
repeat(append(Path, NAME);
    if(NAME == b, done_output(Path));
    hop(all_links); notbelong(NAME, Path))
```

The scenario accumulates each possible path (with all developing independently and in parallel) by collecting names of subsequent nodes during movement through the path, allowing hops to next nodes only if they are not already registered in these growing paths. This scenario will register all paths from g to b including: (g,c,b), (g,f,e,b), (g,f,e,d,b), ...

4.8 Conclusions

Full details on SGL, which continue description of the language features started in Chap. 3 have been given. Summary of SGL syntax and its constructs has been provided and then different types of SGL constants, variables, and rules were presented and explained with their semantics and explanation of use. Elementary examples of programming in SGL have been shown and explained too.

References

1. Sapaty, P.S.: A distributed processing system, European Patent N 0389655, Publ. 10.11.93, European Patent Office
2. Sapaty, P.S.: Mobile Processing in Distributed and Open Environments. John Wiley & Sons, New York (1999)
3. Sapaty, P.S.: Ruling Distributed Dynamic Worlds. John Wiley & Sons, New York (2005)
4. Sapaty, P.S.: Managing Distributed Dynamic Systems with Spatial Grasp Technology. Springer (2017)
5. Sapaty, P.S.: Holistic Analysis and Management of Distributed Social Systems. Springer (2018)
6. Sapaty, P.S.: Complexity in International Security: A Holistic Spatial Approach. Emerald Publishing (2019)
7. Sapaty, P.S.: Symbiosis of Real and Simulated Worlds Under Spatial Grasp Technology. Springer (2021)
8. Sapaty, P.S.: Spatial Grasp as a Model for Space-Based Control and Management Systems. CRC Press (2022)
9. Sapaty, P.S.: The Spatial Grasp Model: Applications and Investigations of Distributed Dynamic Worlds. Emerald Publishing (2023)
10. Sapaty, P.S.: Providing Integrity, Awareness, and Consciousness in Distributed Dynamic Systems. CRC Press (2024)
11. Sapaty P.S.: Spatial management of air and missile defence operations. Math. Mach. Syst. **1** (2023). http://www.immsp.kiev.ua/publications/articles/2023/2023_1/01_23_Sapaty.pdf
12. Sapaty, P.S.: Providing distributed system integrity under Spatial Grasp technology. Math. Mach. Syst. **2** (2023). http://www.immsp.kiev.ua/publications/articles/2023/2023_2/02_23_Sapaty.pdf
13. Sapaty, P.S.: Providing global awareness in distributed dynamic systems. Int. Relat. Dipl. **11**(2), 87–100. https://doi.org/10.17265/2328-2134/2023.02.002. https://www.davidpublisher.com/Public/uploads/Contribute/6486c3d05a6cc.pdf
14. Sapaty P.S., Simulating distributed consciousness with Spatial Grasp model. Math. Mach. Syst. **3** (2023)
15. Sapaty P.S.: Network centrality operations under Spatial Grasp Technology. J. Adv. Artif. Intell. Mach. Learn. **1**(1), 1–11 (2023). https://www.scivisionpub.com/journals/articleinpress-journal-of-advances-in-artificial-intelligence-and-machine-learning
16. Sapaty P.S.: Managing distributed systems with Spatial Grasp patterns. Math. Mach. Syst. **4** (2023). http://www.immsp.kiev.ua/publications/articles/2023/2023_4/04_23_Sapaty.pdf
17. Sapaty P.S.: Spatial networks as models for organoid cultures and brain research. Math. Mach. Syst. **1** (2024)
18. Sapaty, P.S.: Basic network operations under Spatial Grasp model. Sci. Set J. Phys. **3**(1), 01–16 (2024)
19. Sapaty, P.: Spatial Grasp Language (SGL). Adv. Image Video Process. **4**(1) (2016). https://journals.scholarpublishing.org/index.php/AIVP/article/view/1922/1052
20. Sapaty, P.: A brief introduction to the Spatial Grasp Language (SGL). J. Comput. Sci. Syst. Biol. **9**(2) (2016). http://www.omicsonline.org/computer-science-systems-biology.php
21. Sapaty, P.S.: Spatial Grasp Language (SGL) for distributed management and control. J. Robot., Netw. Artif. Life **3**(2) (2016)
22. Sapaty, P.S.: A brief introduction to the Spatial Grasp Language (SGL). J. Comput. Sci. Syst. Biol. **9**(2) (2015). https://www.researchgate.net/publication/307744279_A_Brief_Introduction_to_the_Spatial_Grasp_Language_SGL
23. Sapaty, P.: The language to manage distributed robotized systems. In: Proceedings of Simulation Conference MODS 2007, Zhukin, June, Ukraine (2007)
24. Sapaty P.S.: Mosaic warfare: from philosohpy to model to solution. Math. Mach. Syst. **3** (2019)
25. Sapaty, P.S.: Global network management under Spatial Grasp paradigm. Int. Robot. Autom. J. **6**(3), 134–148 (2020). https://medcraveonline.com/IRATJ/IRATJ-06-00212.pdf

26. Sapaty, P.S.: Global network management under Spatial Grasp paradigm. Global J. Res. Eng.: J General Eng. **20**(5), 58–81. Version 1.0. https://globaljournals.org/GJRE_Volume20/6-Global-Network-Management.pdf
27. Sapaty, P.: Advanced terrestrial and celestial missions under Spatial Grasp Technology. Aeronaut. Aerosp. Open Access J. **4**(3) (2020). https://medcraveonline.com/AAOAJ/AAOAJ-04-00110.pdf
28. Sapaty, P.: Spatial management of distributed social systems. J. Comput. Sci. Res. **2**(3) (2020). https://ojs.bilpublishing.com/index.php/jcsr/article/view/2077/pdf
29. Sapaty, P.: Towards global nanosystems under high-level networking technology, Acta Sci. Comput. Sci. **2**(8) (2020). https://www.actascientific.com/ASCS/pdf/ASCS-02-0051.pdf
30. Sapaty, P.: Symbiosis of distributed simulation and control under Spatial Grasp Technology. SSRG Int. J. Mobile Comput. Appl. (IJMCA) **7**(2) (2020). http://www.internationaljournalssrg.org/IJMCA/2020/Volume7-Issue2/IJMCA-V7I2P101.pdf
31. Sapaty, P.S.: Symbiosis of real and simulated worlds under global awareness and consciousness. In: Abstract, The Science of Consciousness (TSC) Conferences, Tucson, Arizona (2020). https://eagle.sbs.arizona.edu/sc/report_poster_detail.php?abs=3696
32. Sapaty P.S.: Spatial Grasp as a model for space-based control and management systems. Math. Mach. Syst. **1**, 135–138 (2021). http://www.immsp.kiev.ua/publications/articles/2021/2021_1/Sapaty_book_1_2021.pdf
33. Sapaty, P.S.: Managing multiple satellite architectures by Spatial Grasp Technology. Math. Mach. Syst. **1**, 3–16 (2021). http://www.immsp.kiev.ua/publications/eng/2021_1/
34. Sapaty, P.S.: Spatial management of large constellations of small satellites, Math. Mach. Syst. **2** (2021). http://www.immsp.kiev.ua/publications/articles/2021/2021_2/02_21_Sapaty.pdf
35. Sapaty, P.S.: Global management of space debris removal under Spatial Grasp Technology. Acta Sci. Comput. Sci. **3**(7) (2021). https://www.actascientific.com/ASCS/pdf/ASCS-03-0135.pdf
36. Sapaty, P.S.: Space debris removal under Spatial Grasp Technology. Netw. Commun. Technol. **6**(1) (2021). https://www.ccsenet.org/journal/index.php/nct/article/view/0/45486
37. Sapaty, P.S.: Relation of Spatial Grasp paradigm to higher psychological and mental concepts. Acta Sci. Comput. Sci. **4**(12) (2022). https://actascientific.com/ASCS/pdf/ASCS-04-0359.pdf
38. Sapaty, P.S.: Seeing and managing distributed worlds with Spatial Grasp paradigm. Acta Sci. Comput.Sci. **4**(12) (2022). https://actascientific.com/ASCS/pdf/ASCS-04-0365.pdf
39. Sapaty, P.S.: Comprehending distributed worlds with the Spatial Grasp paradigm. Math. Mach. Syst. **1** (2022). http://www.immsp.kiev.ua/publications/articles/2022/2022_1/01_22_Sapaty.pdf
40. Sapaty, P.S.: Wave-WP (World Processing) technology and applications. In: Proceedings of International Conference & Exhibition Defence Technology Asia 2006 (Unmanned vehicle Technology 2006, Mine & Warfare Surveillance Section), Singapore, May 24–25 (2006)
41. Sapaty, P.S.: WAVE-WP (World Processing) technology. Math. Mach. Syst. **3**, 3–17 (2004). ISSN: 1028-9763
42. Sapaty, P.S.: Spatial programming of distributed dynamic worlds in WAVE. In: Presentation at the Special Colloquium "Internet Challenges", Hasso-Plattner-Institut, University of Potsdam, Berlin, Germany, Oct 4, 50p (2002)
43. Sapaty, P.S.: Cooperative exploration of distributed worlds in WAVE. Int. J. Artif. Life Robot. Springer-Verlag Tokyo **4**, 109–118 (2000)
44. Sapaty, P.S.: High-level spatial scenarios in WAVE. In: Proceedings of the International Symposium AROB 5th, Oita, Japan, January, pp. 301–304 (2000)
45. Sapaty, P.S.: Cooperative conquest of distributed worlds in WAVE. In: Proceedings of the Symposium and Exhibition of the Unmanned Systems of the New Millennium, AUVSI'99, Baltimore, MD, July 13–15, 16p (1999)
46. Sapaty, P.S.: Mobile programming in WAVE. Math. Mach. Syst. **1**, 3–31 (1998). Kiev. ISSN: 1028-9763
47. Sapaty, P.S.: Live demonstration of the WAVE system and applications at the Workshop on Mobile Agents and Security 97, Maryland Center for Telecommunications Research, Department of Computer Science and Electrical Engineering, UMBC, October 27–28 (1997)

48. Sapaty, P.S.: WAVE: creating dynamic worlds based on mobile cooperative agents. Dartmouth Workshop on Transportable Agents, Dartmouth College, Hanover, New Hampshire, September (1996)
49. Sapaty, P.S.: Mobile wave technology for distributed knowledge processing in open networks. In: Proceedings of Workshop on New Paradigms in Information Visualization and Manipulation, in conjunction with the Fourth International Conference on Information and Knowledge Management (CIKM'95), Baltimore, Maryland, December (1995)
50. Sapaty, P.S.: A brief introduction to the WAVE language, Report No. 3/93, Faculty of Informatics, University of Karlsruhe (1993)
51. Sapaty, P.S.: The WAVE paradigm, Report No. 17/92, Faculty of Informatics, University of Karlsruhe (1992)
52. Sapaty, P.S.: The WAVE paradigm. In: Proceedings of JICSLP'92 Post-Conference Joint Workshop on Distributed and Parallel Implementations of Logic Programming Systems, Washington, D.C., 13–14 Nov (1992)
53. Sapaty, P.S.: The WAVE model for advanced knowledge processing. In: Ambler, A.P., Agrawal, P., Moore, W.R. (eds.) CAD Accelerators. Elsevier Science Publ. B.V. (1990)
54. Sapaty, P.S.: The WAVE machine project. In: Proceedings of IFIP Workshop on Silicon Architectures for Neural Nets, St. Paul de Vence, France, 28–30 Nov (1990)
55. Sapaty, P.S.: The WAVE model for advanced knowledge processing. Report No. OUEL 1803/89, University of Oxford, England (1989)
56. Sapaty, P.S.: The WAVE-1: a new ideology and language of distributed processing on graphs and networks. Comput. Artif. Intell. **5** (1987)
57. Sapaty, P.S.: The WAVE-0 language as a framework of navigational structures for knowledge bases using semantic networks. In: Proceeding of USSR Academy of Sciences. Technical Cybernetics, No. 5 (1986)
58. Sapaty, S.: A wave language for parallel processing of semantic networks. Comput. Artif. Intell. **5**(4), 289–314 (1986)
59. Sapaty, P.S.: The wave approach to distributed processing of graphs and networks. In: Proceedings of International Working Confernce Knowledge and Vision Processing Systems, Smolenice, November (1986)
60. Sapaty, P.S.: A wave approach to the languages for semantic networks processing. In: Proceedings of International Workshop on Knowledge Representation. Section 1: Artificial Intelligence, Kiev (1984)
61. Sapaty, P.S.: On possibilities of the organization of a direct intercomputer dialogue in ANALYTIC and FORTRAN languages. Publ. No. 74–29, Institute of Cybernetics Press, Kiev (1974)

Chapter 5
Organization and Components of SGL Interpreter

5.1 Introduction

After briefing basics of Spatial Grasp Technology (SGT) and its Spatial Grasp Language (SGL), the chapter explains how SGT can be implemented with communicating SGL interpreter copies easily embedded into any existing distributed systems, internet including, representing altogether powerful networked engines capable of solving any problems in terrestrial and celestial environments.

The chapter is organized as follows. Section 5.2 briefs basics of Spatial Grasp Model and Technology. Section 5.3 shows main components of SGL interpreter, Sect. 5.4 describes data structures of the interpreter, and Sect. 5.5 functional processors of the interpreter. Section 5.6 reveals how track-based automatic command and control works, discussing track components and links with other elements, tracks-based forward grasping, echoing via tracks, and further forward development, via tracks too. Section 5.7 concludes the chapter. References reveal the used SGT-related publications.

5.2 Basics of Spatial Grasp Model and Technology

Only main technology ideas are mentioned below, more can be found in the related patent [1], published books [2–10], latest journal papers [11–18], in numerous previous publications on this paradigm [19–70] and its earlier versions called WAVE [71–91], as well as in previous Chaps. 3 and 4.

Within Spatial Grasp Model and Technology (SGT), a high-level operational scenario expressed in recursive Spatial Grasp Language (SGL), starting in any world point *propagates, covers, and matches the distributed environment in parallel wave-like mode.* as symbolically shown in Fig. 5.1, Such propagation can result in returning

P. S. Sapaty, *Spatial Networking in the United Physical, Virtual, and Mental World*, Studies in Systems, Decision and Control 533,
https://doi.org/10.1007/978-3-031-62154-3_5

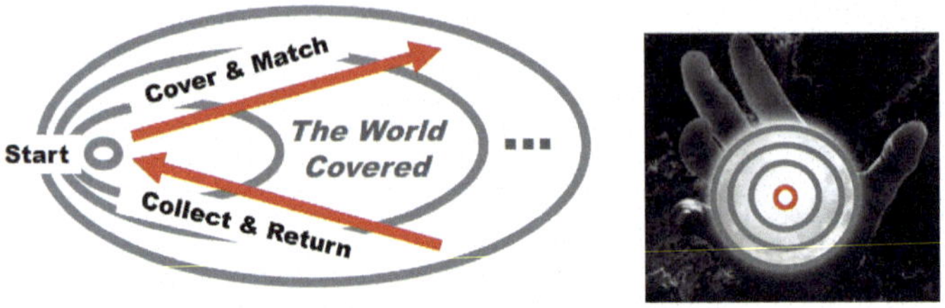

Fig. 5.1 Parallel wavelike world coverage under Spatial grasp model

and analyzing the reached states and data, which may be arbitrarily remote, or used for launching more waves.

SGL allows for *direct space presence and operations* with unlimited powers and parallelism. Its universal recursive organization, with operational scenarios called ***grasp***, can be expressed by a single formula:

$$\boldsymbol{grasp} \rightarrow \boldsymbol{constant}|\boldsymbol{variable}|\boldsymbol{rule}(\{\boldsymbol{grasp},\ \})$$

The ***rule*** expresses certain action, control, description or context accompanied with operands, which can be any ***grasps*** too. The rules, starting in certain points, can organize navigation of the world sequentially, in parallel, or any combinations thereof. They can result in the same application points or cause movement to other world points with obtained results left there or returned.

The *SGL interpreter* consists of specialized modules serving SGL scenarios or their parts happened to be inside this interpreter, also organizing exchanges with other interpreters for distributed SGL scenarios. Communicating SGL interpreters can be in arbitrary number of copies, representing *powerful spatial engines operating without central resources or control*, as in Fig. 5.2. They can effectively work with spatial *graph and network knowledge* of any structure, volume, and distributions.

The following sections provide further details of structure, organization, and operation of SGL interpreter which were briefed in Chap. 3.

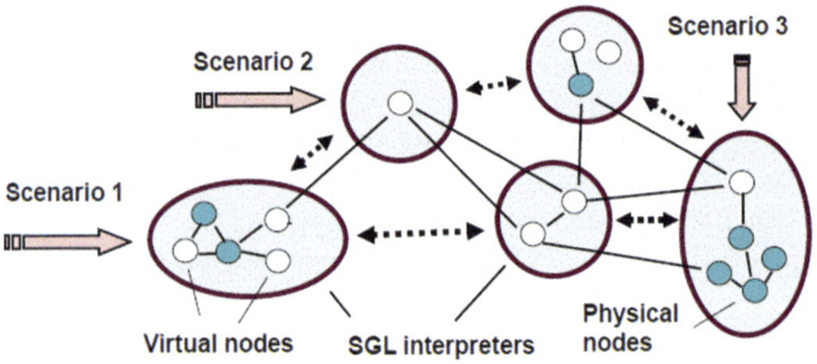

Fig. 5.2 SGL distributed networked interpretation

5.3 Components of SGL Interpreter

The SGL interpreter main components and its general organization are shown in Fig. 5.3.

The interpreter consists of a number of specialized *functional processors* (shown by rectangles) working with and sharing specific data structures. These include: Communication Processor (CP), Control Processor (COP), Navigation Processor (NP), Parser (P), different Operation Processors (OP), and special (external & internal) World Access Unit (WAU) directly manageable from SGL. Main *data structures* (also referred to as *stores*) with which these processors operate (shown by ovals) comprise: Grasps Queue (GQ), Suspended Grasps (SG), Track Forest (TF), Activated Rules (AR), Knowledge Network (NN), Grasps Identities (GI), Heritable Variables (HV), Fontal Variables (FV), Nodal Variables (NV), Environmental Variables (EV), Global Variables (GV), Incoming Queue (IQ), and Outgoing Queue (OQ).

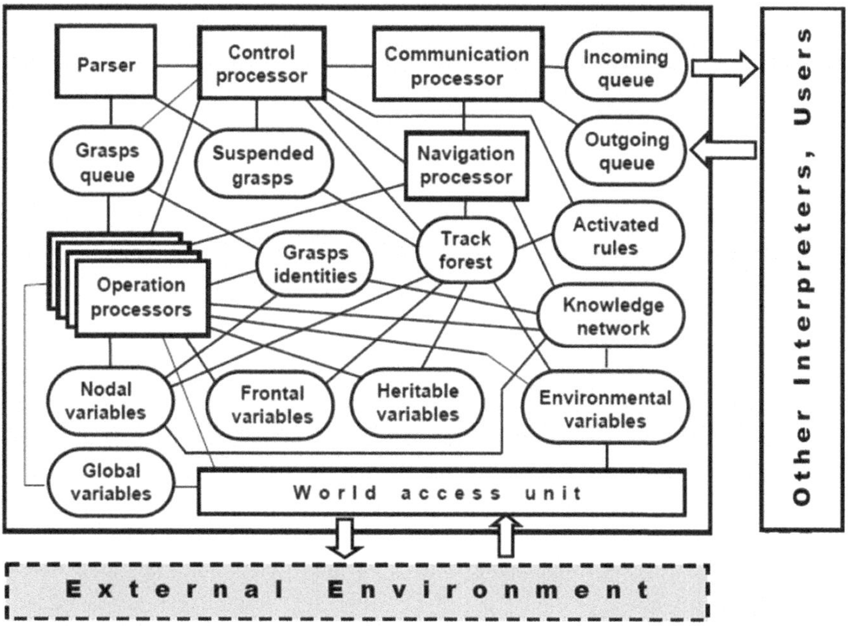

Fig. 5.3 SGL interpreter main components and their interactions

5.4 Data Structures of the Interpreter

SGL interpretation network generally serves multiple scenarios or their parallel branches simultaneously navigating the distributed world, which can cooperate or compete with each other. Each interpreter can support and process multiple SGL scenario code which happens to be in its responsibility at different moments of time. This is reflected in the interpreter's basic data structures which hold permanent, persistent, and/or temporary information about the navigated and processed worlds.

Grasps Queue. The Grasps Queue (GQ) keeps multiple scenario fragments (syntactically represented as *grasps*) which are parsed and ready for execution in this particular interpreter. Independent from each other and queued by the priority or arrival time, they can be processed sequentially or in parallel (if the latter capabilities exist in the individual interpreter). Partially executed grasps (say, needing additional execution functionalities) can be returned to the same GQ for further processing.

Suspended Grasps. Suspended Grasps (SG) structure keeps deferred parts of SGL scenarios which will be parsed in detail and executed in this interpreter or forwarded to other interpreters when proper conditions are met, control states and/or data availability including. The sequence of recorded grasps in SG *does not represent a queue*, as the time and order of their invocation depend on situations within distributed control of multiple operations in the current and other interpreters.

Track Forest. Track Forest (TF) is a special self-optimizing structure reflecting *dynamic history* of spatial evolution of SGL scenarios, allowing us to automatically coordinate and control multiple distributed processes with making decisions at different levels. It preserves *integrity* of the whole set of parallel and distributed processes as a global goal-driven system, also supports existence and controls lifetime of different types of spatial variables described earlier in this chapter. TF spans throughout the navigated world with its interlinked parts kept in different interpreters while forming altogether a seamless spatial organizational, command and control infrastructure. Its work will be detailed later in this chapter.

Activated rules. Activated Rules (AR) store represents rules that have been activated and continue working with different scenarios or their branches. Some rules may be waiting for their operands to be completed (say, for obtaining data to be processed or the echoed control states to make decisions). These operands may be arbitrary grasps evolving in the same or in other interpreters and integrated within SGL scenarios by means of track trees in TF. Rules registered in AR are associated with certain TF nodes and the latter, in their turn, with related grasps in GQ and SG, which may represent the rule's operand scenarios currently in active, passive, or suspended states. The rules in AR may be of any kinds—from simple ones to their aggregates with personal parameters, which may need to be processed first to obtain the values needed. The data-processing rules with all operands completed (i.e. which can be directly executed) are placed into GQ. Rules in AR may form a hierarchy covering via tracks more than one interpreter and operating as an integral system with the help of distributed TF structures.

Knowledge Network. The Knowledge Network (KN) store keeps interlinked data on parts of virtual, physical, execution, or combined worlds associated with the current interpreter, this including named or unnamed nodes (which may have physical coordinates) and named links connecting them. These parts may belong to different applications with their specific world domains, which are navigated and processed by different SGL scenarios, where the latter may cooperate, compete, or just ignore each other. Nodes of KN fragments in the current interpreter may have semantic links with nodes located in other interpreters thus forming integral distributed information infrastructures covering application areas. Any node navigated or created by the current interpreter and belonging to PW, VW, EW or their combinations is immediately registered in the KN store. Except unnamed pure PW nodes, which automatically disappear as soon as all activities related to them cease, all other nodes and relations between them in KN remain regardless of being or not being accessed by SGL scenarios, unless explicitly deleted by special rules.

Grasps Identities. The Grasps Identities (GI) are keeping personal colors, or identities, of scenarios that have been currently processed in this interpreter (in an active or suspended way), directing them to the resources related to these scenarios, like nodal variables and KN elements. These can be accessed and shared by different scenarios or their branches with the same identities (or which are aware of these identities while being themselves of other personal colors). This coloring of distributed resources (potentially hierarchical) can be effectively used for protecting own information from unauthorized access while allowing users to work cooperatively with other users on distributed resources from different application areas.

Heritable Variables. The Heritable Variables (HV) store keeps all heritable variables with their contents created by SGL scenarios within the current interpreter, these variables being linked to the Track Forest nodes within which they originated. The heritable variables can be accessed by all processes stemming from these track nodes, which may be developing in this or other interpreters. This access is achievable via propagation through the distributed track trees (and between interpreters, if needed) in both read and write mode.

Frontal Variables. The Frontal Variables (FV) store holds all frontal variables with their contents registered at the current moment of time within the current interpreter. These variables are linked with the fringe nodes of evolving track trees reflecting space–time evolution of generally hierarchical and parallel processes, being always associated with the latest steps of the interpreted scenarios (having moved to them from the preceding steps). Frontal variables are a sole property of these latest scenario stages; they will be replicated if the scenario splits into branches, becoming a non-sharable property of each new branch. If scenarios have to move to other interpreters, frontal variables will be moving with them too, withdrawing themselves from FV store in the current interpreter.

Nodal Variables. Nodal Variables (NV) store keeps all nodal variables with their contents which are linked to nodes in the KN store being their temporary property until the scenarios that created them remain alive in the distributed space (not necessarily in the current interpreter only). For that reason, they are also connected with corresponding nodes in TF store, as after the full completion of scenarios all their track structures, which may spread between interpreters, automatically disappear together with heritable and nodal variables linked to them. Nodal variables are also connected to grasps identities in the GI store through which they can be accessed and shared by all SGL scenarios having these identities (or knowing them).

Environmental Variables. Environmental Variables (EV) store keeps special variables with reserved names, which allow us to control and work directly with different space and time features of the worlds created, navigated, and processed, also with internal parameters of the interpreter. Most of these variables instead of having their own contents (like the previous heritable, frontal, and nodal ones) are referring to parameters of other structures or the worlds currently navigated (like those registered in KN) by evolving SGL scenarios. They may also be accessing special hardware or software represented by the World Access Unit (like different timing devices, sensors, and channels), or allowing us to directly communicate with local human users or various external devices. Most of these variables are classified as stationary but some may behave like frontal ones (e.g. scenario colors) temporarily linking with nodes in TF store on their spatial move.

Global Variables. Global Variables (GV) store keeps information on global variables, being the most expensive ones, which can be simultaneously used and shared by any scenarios or their different branches, and at any time. Their management is beyond traditional distributed track-supported interpreter organization and can be implemented with the help of Word Access Unit capabilities allowing for direct access of global external stores (or other systems) keeping their contents. In many cases, however, the effect of global variables can be achieved by the use of heritable variables if the latter are declared ahead of the scenario development, but only for individual scenarios, not sets of them where they are independent from each other.

Incoming Queue. Incoming Queue (IQ) accepts and stores incoming messages, which may be complete SGL scenarios from users or their parts coming from other interpreters, to be executed in the current interpreter and, possibly, continued in other interpreters afterwards. The messages may also be remotely obtained control states or data to be analyzed and processed in this interpreter or forwarded further. The incoming messages may have different priorities, with control messages being of highest urgency, so the procedures for processing IQ elements must take into account their both arrival order and importance.

Outgoing Queue. The Outgoing Queue (OQ) store accumulates obtained results to be passed to the directly connected users, parts of SGL scenarios with accompanying intermediate data to be forwarded to other interpreters for further consideration, control states and data to be returned to other interpreters on their request, as well as direct commands to users or neighboring interpreters. As for the IQ, the procedures of processing OQ messages may combine the first-come-first-served strategy with superiority of command, control and emergency messages exchanged with other parts of the distributed system.

5.5 Functional Processors of the Interpreter

Functional Processors carry different SGL interpretation loads, serve interpreter's data structures, and communicate with other interpreters and external users. They also provide system integrity and overall command and control of local and global operations in potentially distributed and dynamic environments.

Communication Processor. Communication Processor (CP) receives SGL scenarios or their parts, service and control messages, and requested remote data from the external world or other interpreters, classifying them accordingly and sending to other processors for execution. CP also allows the current interpreter to serve as a transit node in exchanges between other interpreters. The incoming and outgoing messages are optimized by CP for efficient communication with other interpreters, also for processing within the current interpreter. CP directly operates with IQ and OQ, and communicates with Control Processor and Navigation Processor via internal buffers, allowing them to operate asynchronously and in parallel with each other.

Parser. Parser (P) carries out syntactic analysis of the scenarios or their parts, extracts control rules in them, decomposes SGL strings into ready for execution elementary grasps (with clarified operands) and the remainders to be suspended for further consideration. The latter can take place when proper conditions are met or values of operands for rules (which themselves may be arbitrary grasps) are finally obtained. Parser also optimizes and compresses SGL code for its further processing (like removing blanks and substituting rule names and special words by short abbreviations). Parser directly operates with GQ and SG, also communicates with Control Processor via internal buffer allowing the two processors to work asynchronously and in parallel with each other.

Operation Processors. Operation Processors (OP) unit performs basic analysis and operation procedures over information units and physical matter (or physical objects), expressed by rules. It works directly with GQ, GI, NV, FV, HV, EV, GV, and KN stores. OP also directly communicates with Control Processor, Navigation Processor, and World Access Unit. In case of compound rules, the rest of SGL string taken from GQ for an operation and partially processed in OP can be returned to GQ for a continued execution by other operation processors.

Navigation Processor. Navigation Processor (NP) specializes in creation and navigation of network-structured data, operating directly with TF and KN stores. These data networks (more persistent in KN and temporary in TF) may both be distributed between interpreters. In this case NP performs network navigation in the current interpreter while transferring control and orders to NPs in other interpreters when reaching the network boundaries in the current interpreter, for a continued navigation. As regards the history tracks in TF, they may be optimized (by substituting node sets with single nodes when history details become redundant) or removed partially or completely by NP upon termination of SGL scenarios or their parts. NP cooperates directly with Control Processor, Communication Processor, and Operation Processors.

Control Processor. Control Processor (COP) provides local and global control and coordination of sequential and parallel processes within the interpreter including classification and forwarding of different types of messages within and between interpreters and decomposition of SGL strings while sending them for execution or suspending till proper conditions met. COP also provides interpretation of all control rules of SGL, supporting the distributed command and control hierarchy based on history tracks, which may spread to and cover other interpreters. COP directly operates on GQ, SG, TF, and AR stores. It also cooperates with CP, P, NP, and OP, supervising collective work of these processors within different SGL interpretation procedures. COP plays the key role in organization and support of track-based distributed management and control.

World Access Unit. World Access Unit (WAU) offers an extension to the interpreter's main functionality for interaction with external (and also interpreter's internal) physical and virtual environments, also an interface for integration with other systems like internet, robotic equipment, and all those working with physical matter or objects, global variables too. WAU can be accessed via environmental and global variables as well as directly from processors in OP.

5.6 Tracks-Based Automatic Command and Control

As both *backbone and nerve system* of the distributed interpreter, its hierarchical spatial track system dynamically spans the worlds in which SGL scenarios evolve, providing automatic control of multiple distributed processes. Its part related to the current interpreter is kept in Track Forest store which is interlinked with similar parts in other interpreters, forming altogether *global control coverage*. Self-optimizing in parallel echo processes, this (generally forest-like) distributed track structure provides hierarchical command and control as well as remote data and code access. It also supports spatial variables and merges distributed control states for making decisions at different organizational levels. The track infrastructure can be automatically distributed between different doers during scenario spreading in distributed environments.

- **Track components and links with other elements**

These being as follows:

- *Track nodes* reflecting scenario progress points (or *props*)—the stages through which spatial scenarios evolve and form their development history.
- *Track links* providing transition, succession between consecutive props.
- *Heritable variables* as sole track node properties capable of being accessed by all operations related to the current and subsequent props.
- *Frontal variables* accompanying the scenario evolution and being associated at any time with the latest, fringe track nodes.

- *World nodes* belonging to virtual, physical, executive or combined worlds navigated by the scenario and registered in KN store; these being generally linked with sets of track nodes (as sequences or trees rather than single track nodes).
- *Nodal variables* associated with the world nodes created and/or navigated by SGL scenarios. These variables are also linked with particular track nodes under which they were formed. If these track nodes are removed in the tracks cleaning process caused by termination of scenarios or their parts, these variables will be removed too. Nodal variables will also be deleted if the world nodes to which they belong cease to exist (say, after explicit removal in the scenarios using them).
- *Activated rules* linked with certain track nodes. They start their influence within corresponding props and may use the subsequent track tree (to its full depth) emanating from these nodes for managing and supervising of the rule-related forward and echo operations.
- *Suspended grasps* associated with activated rules and connected to the same track nodes as themselves. These grasps will be subsequently launched by these rules after proper conditions are met, using track infrastructure emanating from this node for their forwarding.

Some of the main track components are shown in Fig. 5.4, which will be also used in the subsequent figures.

- **Tracks-based forward grasping**

In the forward SGL scenario process, as in Fig. 5.5, the next steps of scenario development can be considered as staying with the same track nodes or forming new track nodes connected to the previous ones by track links. Except reflecting history of scenario evolution, this growing track tree, as already mentioned, is supporting heritable, nodal and frontal variables as well as activated rules and suspended grasps (all being associated with proper tack nodes).

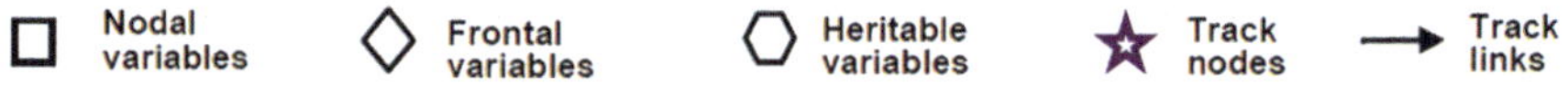

Fig. 5.4 Main track components

Fig. 5.5 Forward world grasping

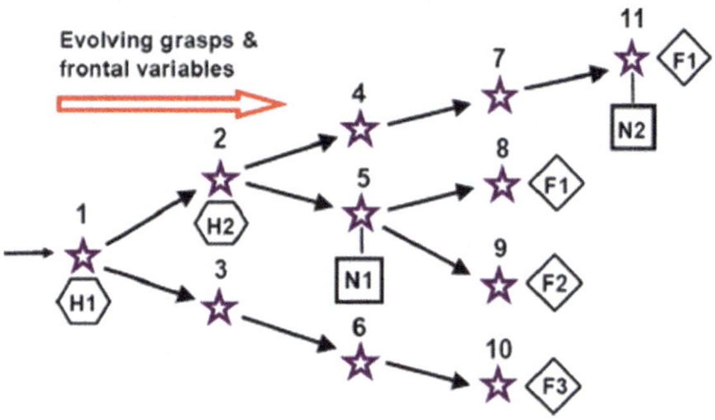

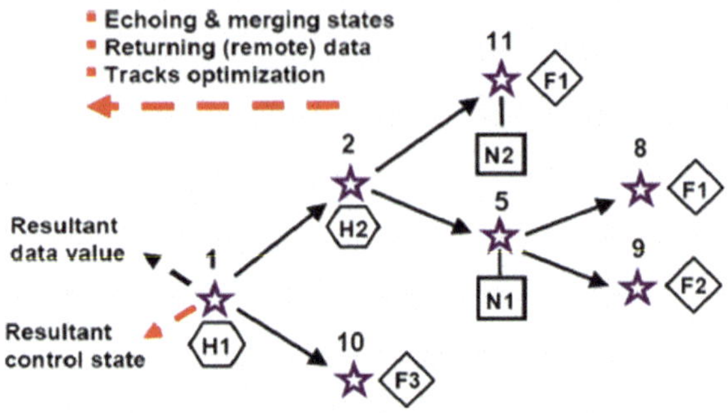

Fig. 5.6 Echoing and tracks optimization

Track nodes are directly associated with particular world nodes at which they appear (following the SGT ideology where all SGL processes are always linked with certain world points where they take place) and this association is inherited by all subsequent track nodes unless they fall into alliance with other world nodes. The latter, in turn, will be inherited by the subsequent track nodes unless shifting to responsibility of other world nodes, and so on. The track infrastructure can be automatically distributed between different doers during scenario spreading in distributed environments.

- **Echoing via tracks**

After completing the forward stage of SGL scenario discussed above, the track system can return to the starting track node the generalized control state based on termination states in all fringe props, as in Fig. 5.6, also marking the passed track links with the states returned via them.

The states generalization process is based on priority of control states (from strongest to weakest as: *fatal*, *thru*, *done*, and *fail*). The track system, on the request of higher-level scenario rules, can also collect local data obtained at its fringe props and merge them into a resultant list of values echoed to the starting prop. The track echoing process also optimizes the track system for its further use, for example, by deleting already used and not needed any more items associated with it, as in Fig. 5.6 in comparison with Fig. 5.5.

- **Further forward development via tracks**

The echo-modified and optimized track system after the previous scenario stage completion can route further grasps to the world positions reached by the previous grasps and defined by fringe track nodes having state *thru*, as in Fig. 5.7. Heritable variables created in certain track nodes can be accessed from the subsequent nodes in the track system for both reading and writing operations, and at any depth of the evolving track tree, by following links between track nodes in both directions.

More details and examples related to the organization of track system effectively supporting distributed SGL interpretation can be found in the existing publications, and particularly in [2–10].

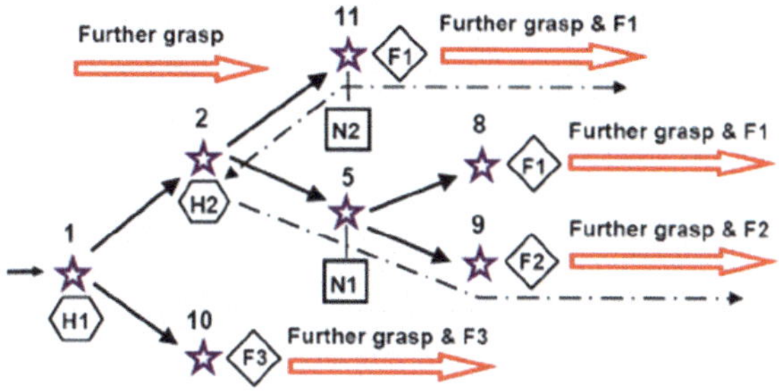

Fig. 5.7 Further world grasping

5.7 Conclusions

General ideas of the mobile Spatial Grasp Technology (SGT) and its basic Spatial Grasp Language (SGL) have been summarized. The structure of SGL interpreter was described, with its communicating copies forming powerful parallel networked engines covering arbitrary large environments. The purpose and organization of the spatial tracking system of the interpreter effectively supporting spatial forward–backward capabilities of SGL have been explained.

References

1. Sapaty, P.S.: A distributed processing system. European Patent N 0389655, Publ. 10.11.93, European Patent Office
2. Sapaty, P.S.: Mobile Processing in Distributed and Open Environments. Wiley, New York (1999)
3. Sapaty, P.S.: Ruling Distributed Dynamic Worlds. Wiley, New York (2005)
4. Sapaty, P.S.: Managing Distributed Dynamic Systems with Spatial Grasp Technology. Springer, Berlin (2017)
5. Sapaty, P.S.: Holistic Analysis and Management of Distributed Social Systems. Springer, Berlin (2018)
6. Sapaty, P.S.: Complexity in International Security: A Holistic Spatial Approach. Emerald Publishing (2019)
7. Sapaty, P.S.: Symbiosis of Real and Simulated Worlds under Spatial Grasp Technology. Springer, Berlin (2021)
8. Sapaty, P.S.: Spatial Grasp as a Model for Space-based Control and Management Systems. CRC Press (2022)
9. Sapaty, P.S.: The Spatial Grasp Model: Applications and Investigations of Distributed Dynamic Worlds. Emerald Publishing (2023)
10. Sapaty, P.S.: Providing Integrity, Awareness, and Consciousness in Distributed Dynamic Systems, CRC Press (2024)
11. Sapaty, P.S.: Spatial management of air and missile defence operations. Math. Mach. Syst. 1 (2023). http://www.immsp.kiev.ua/publications/articles/2023/2023_1/01_23_Sapaty.pdf
12. Sapaty, P.S.: Providing distributed system integrity under spatial grasp technology. Math. Mach. Syst. 2 (2023). http://www.immsp.kiev.ua/publications/articles/2023/2023_2/02_23_Sapaty.pdf

13. Sapaty, P.S.: Providing global awareness in distributed dynamic systems. Int. Relat. Dipl. **11**(2), 87–100 (2023). https://doi.org/10.17265/2328-2134/2023.02.002. https://www.davidpublisher.com/Public/uploads/Contribute/6486c3d05a6cc.pdf
14. Sapaty, P.S., Simulating Distributed Consciousness with Spatial Grasp Model, Mathematical machines and systems. N 3, 2023.
15. Sapaty, P.S.: Network centrality operations under spatial grasp technology. J. Adv. Artif. Intell. Mach. Learn. **1**(1), 1–11 (2023). https://www.scivisionpub.com/journals/articleinpress-journal-of-advances-in-artificial-intelligence-and-machine-learning
16. Sapaty, P.S.: Managing distributed systems with spatial grasp patterns. Math. Mach. Syst. 4 (2023). http://www.immsp.kiev.ua/publications/articles/2023/2023_4/04_23_Sapaty.pdf
17. Sapaty, P.S.: Spatial networks as models for organoid cultures and brain research. Math. Mach. Syst. 1 (2024)
18. Sapaty, P.S.: Basic network operations under spatial grasp model. Sci. Set. J. Phys. **3**(1), 01–16 (2024)
19. Sapaty, P.S.: Distributed artificial brain for collectively behaving mobile robots. In: Proceedings of Symposium and Exhibition Unmanned Systems 2001, Baltimore, MD, Jul 31–Aug 2, 2001
20. Sapaty, P.S.: Crisis management with distributed processing technology. Int. Trans. Syst. Sci. Appl. **1**(1), 81–92 (2006). ISSN 1751-1461.
21. Sapaty, P.S.: Global management of distributed EW-related system. In: Proceedings of International Conference Electronic Warfare: Operations & Systems 2007, Thistle
22. Sapaty, P.S.: Distributed technology for global dominance. In: Proceedings of International Conference Defense Transformation and Net-Centric Systems 2008, as part of the SPIE Defense and Security Symposium, World Center Marriott Resort and Convention Center, Orlando, FL, USA, 16–20 Mar 2008 (Also: Proceedings of SPIE—Volume 6981, Defense Transformation and Net-Centric Systems 2008, Raja Suresh, Editor, 69810T, Apr. 3, 2008)
23. Sapaty, P.S.: Distributed technology for global dominance. In: Keynote Lecture, Proceedings of the Fifth International Conference in Control, Automation and Robotics ICINCO 2008, The Conference Proceedings, Funchal, Madeira, Portugal, 11–15 May 2008
24. Sapaty, P.S.: Human-robotic teaming: a compromised solution. In: AUVSI's Unmanned Systems North America 2008, San Diego, USA, 10–12 June 2008
25. Sapaty, P.S.: High-level communication protocol for dynamically networked battlefields. In: Proceedings of International Conference Tactical Communications 2009 (Situational Awareness and Operational Effectiveness in the Last Tactical Mile), One Whitehall Place, Whitehall Suite and Reception, London, UK (2009)
26. Sapaty, P.S.: Distributed capability for battlespace dominance. In: Electronic Warfare 2009 Conference and Exhibition, Novotel London West Hotel and Conference Center, London, 14–15 May 2009
27. Sapaty, P.S.: Providing spatial integrity for distributed unmanned systems. In: Proceedings of 6th International Conference in Control, Automation and Robotics ICINCO 2009, Milan, Italy (2009)
28. Sapaty, P.S.: Distributed technology for global control. In: Lecture Notes in Electrical Engineering, Informatics in Control, Automation and Robotics, vol. 37. Springer, Berlin (2009)
29. Sapaty, P.S.: Gestalt-based integrity of distributed networked systems. In: SPIE Europe Security + Defence, bcc Berliner Congress Centre, Berlin, Germany (2009)
30. Sapaty, P.S.: Remote control of open groups of remote sensors. In: Proceedings of SPIE Europe Security + Defence, Berlin, Germany (2009).
31. Sapaty, P.S.: Tactical communications in advanced systems for asymmetric operations. In: Proceedings of Tactical Communications 2010, CCT Venues, Canary Wharf, London UK, 28–30 Apr 2010
32. Sapaty, P.S.: High-level technology to manage distributed robotized systems. In: Proceedings of Military Robotics 2010, Jolly St Ermins, London, UK, 25–27 May 2010
33. Sapaty, P.S.: Emerging asymmetric threats, Q&A session. In: Tactical Communications 2010, CCT Venues, Canary Wharf, London UK, 28–30 Apr 2010

34. Sapaty, P.S.: High-level organization and management of directed energy systems. In: Proceedings of Directed Energy Weapons 2010, CCT, Canary Wharf, London UK, 25–26, Mar 2010
35. Sapaty, P.S.: Unified transition to robotized armies with spatial grasp technology. In: International Summit Military Robotics, London, United Kingdom, 12–13 Nov 2012
36. Sapaty, P.S.: Distributed air and missile defense with spatial grasp technology. Intell. Control Autom. Sci. Res. **3**(2), 117–131 (2012)
37. Sapaty, P.S.: Global electronic dominance. In: 12th International Fighter Symposium, 6th–8th November 2012, Grand Connaught Rooms, London, UK
38. Sapaty, P.S.: Providing global awareness in distributed dynamic environments. In: International Summit ISR, London, 16–18 Apr 2013
39. Sapaty, P.S.: Ruling distributed dynamic worlds with spatial grasp technology. In: Tutorial at the International Science and Information Conference 2013 (SAI), London, UK, 7–9 Oct 2013
40. Sapaty, P.S.: Night vision under advanced spatial intelligence: a key to battlefield dominance. In: International Summit Night Vision 2013, London, 4–6 June 2013
41. Sapaty, P.S.: Integration of ISR with advanced command and control for critical mission applications. In: SMi's ISR Conference, Holiday Inn Regents Park, London, 7–8 Apr 2014
42. Sapaty, P.S.: Unified transition to cooperative unmanned systems under spatial grasp paradigm. Int. J. Trans. Netw. Commun. **2**(2) (2014)
43. Sapaty, P.S.: From manned to smart unmanned systems: a unified transition. In: SMi's Military Robotics, Holiday Inn Regents Park London, 21–22 May 2014
44. Sapaty, P.S.: Unified transition to cooperative unmanned systems under spatial grasp paradigm. In: 19th International Command and Control Research and Technology Symposium, Alexandria, Virginia, 16–19 June 2014
45. Sapaty, P.S.: Distributed human terrain operations for solving national and international problems. Int. Relat. Dipl. **2**(9) (2014)
46. Sapaty, P.S.: Providing over-operability of advanced ISR systems by a high-level networking technology. In: SMI's Airborne ISR, Holiday Inn Kensington Forum, London, United Kingdom, 26–27 Oct 2015
47. Sapaty, P.S.: Distributed missile defence with spatial grasp technology. In: SMi's Military Space, Holiday Inn Regents Park London, 4–5 Mar 2015
48. Sapaty, P.S.: Military robotics: latest trends and spatial grasp solutions. Int. J. Adv. Res. Artif. Intell. **4**(4) (2015)
49. Sapaty, P.S.: A brief introduction to the spatial grasp language (SGL). J. Comput. Sci. Syst. Biol. **9**(02) (2015). https://www.researchgate.net/publication/307744279_A_Brief_Introduction_to_the_Spatial_Grasp_Language_SGL
50. Sapaty, P.S.: Organization of advanced ISR systems by high-level networking technology. Math. Mach. Syst. 1 (2016)
51. Sapaty, P.S.: Towards massively robotized systems under spatial grasp technology. J. Comput. Sci. Syst. Biol. **9**(1) (2016)
52. Sapaty, P.S.: Towards wholeness and integrity of distributed dynamic systems. J. Comput. Sci. Syst. Biol. **9**(3) (2016)
53. Sapaty, P.S.: Towards global goal orientation, robustness and integrity of distributed dynamic systems. J. Int. Relat. Dipl. **4**(6) (2016)
54. Sapaty, P.: Spatial grasp language (SGL). Adv. Image Video Process. **4**(1) (2016). https://journals.scholarpublishing.org/index.php/AIVP/article/view/1922/1052
55. Sapaty, P.S.: Mosaic warfare: from philosophy to model to solution. Math. Mach. Syst. 3 (2019)
56. Sapaty, P.S.: Global network management under spatial grasp paradigm. Int. Robot. Autom. J. **6**(3), 134–148 (2020). https://medcraveonline.com/IRATJ/IRATJ-06-00212.pdf
57. Sapaty, P.S.: Global network management under spatial grasp paradigm. Glob. J. Res. Eng.: J Gen. Eng. **20**(5), 58–81 (2020); Version 1.0. https://globaljournals.org/GJRE_Volume20/6-Global-Network-Management.pdf
58. Sapaty, P.: Advanced terrestrial and celestial missions under spatial grasp technology. Aeronaut. Aerosp. Open Access J. **4** (3) (2020). https://medcraveonline.com/AAOAJ/AAOAJ-04-00110.pdf

59. Sapaty, P.: Spatial management of distributed social systems. J. Comput. Sci. Res. **2**(3) (2020). https://ojs.bilpublishing.com/index.php/jcsr/article/view/2077/pdf
60. Sapaty, P.: Towards global nanosystems under high-level networking technology. Acta Sci. Comput. Sci. **2**(8) (2020). https://www.actascientific.com/ASCS/pdf/ASCS-02-0051.pdf
61. Sapaty, P.: Symbiosis of distributed simulation and control under spatial grasp technology. SSRG Int. J. Mob. Comput. Appl. (IJMCA) **7**(2) (2020). http://www.internationaljournalssrg.org/IJMCA/2020/Volume7-Issue2/IJMCA-V7I2P101.pdf
62. Sapaty, P.S.: Symbiosis of real and simulated worlds under global awareness and consciousness. In: Abstract, The Science of Consciousness (TSC) Conferences, Tucson, Arizona (2020). https://eagle.sbs.arizona.edu/sc/report_poster_detail.php?abs=3696
63. Sapaty, P.S.: Spatial grasp as a model for space-based control and management systems. Math. Mach. Syst. **1**, 135–138 (2021). http://www.immsp.kiev.ua/publications/articles/2021/2021_1/Sapaty_book_1_2021.pdf
64. Sapaty, P.S.: Managing multiple satellite architectures by spatial grasp technology. Math. Mach. Syst. 1 3–16. http://www.immsp.kiev.ua/publications/eng/2021_1/
65. Sapaty, P.S.: Spatial management of large constellations of small satellites. Math. Mach. Syst. 2 (2021). http://www.immsp.kiev.ua/publications/articles/2021/2021_2/02_21_Sapaty.pdf
66. Sapaty, P.S.: Global management of space debris removal under spatial grasp technology. Acta Sci. Comput. Sci. **3**(7) (2021). https://www.actascientific.com/ASCS/pdf/ASCS-03-0135.pdf
67. Sapaty, P.S.: Space debris removal under spatial grasp technology. Netw. Commun. Technol. **6**(1) (2021). https://www.ccsenet.org/journal/index.php/nct/article/view/0/45486
68. Sapaty, P.S.: Relation of spatial grasp paradigm to higher psychological and mental concepts. Acta Sci. Comput. Sci. **4**(12) (2022). https://actascientific.com/ASCS/pdf/ASCS-04-0359.pdf
69. Sapaty, P.S.: Seeing and managing distributed worlds with spatial grasp paradigm. Acta Sci. Comput. Sci. **4**(12) (2022). https://actascientific.com/ASCS/pdf/ASCS-04-0365.pdf
70. Sapaty, P.S.: Comprehending distributed worlds with the spatial grasp paradigm. Math. Mach. Syst. 1 (2022). http://www.immsp.kiev.ua/publications/articles/2022/2022_1/01_22_Sapaty.pdf
71. Sapaty, P.S.: Wave-WP (World Processing) technology and applications. In: Proceeding of International Conference & Exhibition DEFENCE TECHNOLOGY ASIA 2006 (Unmanned vehicle Technology 2006, Mine & Warfare Surveillance Section), Singapore, May 24–25, 2006
72. Sapaty, P.S.: WAVE-WP (World Processing) technology. Math. Mach. Syst. 3–17 (2004). ISSN: 1028-9763, No. 3
73. Sapaty, P.S.: Spatial programming of distributed dynamic worlds in WAVE. In: Presentation at the special colloquium "Internet Challenges", Hasso-Plattner-Institut, University of Potsdam, Berlin, Germany, Oct. 4, 2002, 50p
74. Sapaty, P.S.: Cooperative exploration of distributed worlds in WAVE. Int. J. Artif. Life Robot. Springer-Verlag Tokyo **4**, 109–118 (2000)
75. Sapaty, P.S.: High-level spatial scenarios in WAVE. In: Proceedings of the International Symposium AROB 5th, Oita, Japan, January 2000, pp. 301–304
76. Sapaty, P.S.: Cooperative conquest of distributed worlds in WAVE. In: Proceedings of the Symposium and Exhibition of the Unmanned Systems of the New Millennium, AUVSI'99, Baltimore, MD, July 13–15, 1999, 16p.
77. Sapaty, P.S.: Mobile Programming in WAVE. In: Mathematical Machines and Systems, ISSN: 1028–9763, No. 1, January-March 1998, Kiev, pp. 3–31.
78. Sapaty, P.S.: Live Demonstration of the WAVE System and Applications at the Workshop on Mobile Agents and Security 97, Maryland Center for Telecommunications Research, Department of Computer Science and Electrical Engineering, UMBC, October 27–28, 1997
79. Sapaty, P.S.: WAVE: creating dynamic worlds based on mobile cooperative agents. In: Dartmouth Workshop on Transportable Agents, Dartmouth College, Hanover, New Hampshire, September 1996
80. Sapaty, P.S.: Mobile wave technology for distributed knowledge processing in open networks. In: Proceedings of Workshop on New Paradigms in Information Visualization and Manipulation, in conjunction with the Fourth International Conference on Information and Knowledge Management (CIKM'95), Baltimore, Maryland, December 1995

81. Sapaty, P.S.: A brief introduction to the WAVE language. Report No. 3/93, Faculty of Informatics, University of Karlsruhe, 1993
82. Sapaty, P.S.: The WAVE paradigm. Report No. 17/92, Faculty of Informatics, University of Karlsruhe, 1992
83. Sapaty, P.S.: The WAVE paradigm. In: Proceedings of JICSLP'92 Post-Conference Joint Workshop on Distributed and Parallel Implementations of Logic Programming Systems, Washington, D.C., Nov.13–14, 1992
84. Sapaty, P.S.: The WAVE model for advanced knowledge processing. In: Ambler, A.P., Agrawal, P., Moore, W.R. (Eds.), CAD Accelerators. Elsevier Science Publisher B.V. (1990)
85. Sapaty, P.S.: The WAVE machine project. In: Proceeding of IFIP Workshop on Silicon Architectures for Neural Nets, St. Paul de Vence, France, November 28–30, 1990
86. Sapaty, P.S.: The WAVE model for advanced knowledge processing. In: Report No. OUEL 1803/89, University of Oxford, England, 1989
87. Sapaty, P.S.: The WAVE-1: a new ideology and language of distributed processing on graphs and networks. In: Computers and Artificial Intelligence, No.5, 1987
88. Sapaty, P.S.: The WAVE-0 language as a framework of navigational structures for knowledge bases using semantic networks. In: Proceedings of USSR Academy of Sciences. Technical Cybernetics, No. 5, 1986
89. Sapaty, S.: A wave language for parallel processing of semantic networks. Comput. Artif. Intell. **5**(4), 289–314 (1986)
90. Sapaty, P.S.: The wave approach to distributed processing of graphs and networks. In: Proceedings of International Working Conference Knowledge and Vision Processing Systems, Smolenice, November 1986
91. Sapaty, P.S.: A wave approach to the languages for semantic networks processing. In: Proceedings of International Workshop on Knowledge Representation. Section 1: Artificial Intelligence, Kiev, 1984

Chapter 6
Basic Network Operations in Different Worlds Under SGT

6.1 Introduction

Networks as models and operations are widely used in practically every area of human activity. In Chap. 2 we analyzed and reviewed very different types and areas of networking, which included: *Transport Networks* [1–6], *Communication Networks* [7–13], *Social Networks* [14–18], *Battle and Military Networks* [19–23], *Economic Networks* [24–29], *Distribution Networks* [30–34], *Virtual Networks* [35–39], *Neural Networks* [40–44], *Psychological Networks* [45–49], and *Criminal Networks* [50–55]. The main goal of this chapter is to investigate and analyze potential and practical suitability and applicability of the developed and patented Spatial Grasp Technology (SGT) and its basic Spatial Grasp Language (SGL) [56–73], for expressing and solving basic networking problems, and especially in large distributed environments. The rest of the Chapter is organized as follows.

Section 6.2 provides a review of publications on main network and graph operations. Section 6.3 briefs the developed Spatial Grasp Model and Technology, described in detail in Chaps. 3–5. Section 6.4 describes elementary networking operations in SGL, including creating individual nodes and links, network topology formation, and nodes and links removal. Section 6.5 is devoted to paths in networks and describes finding and collecting any and all paths between certain nodes. Section 6.6 describes creation of different types of spanning trees from a node, like breadth-first and depth-first, also shortest path tree to all other nodes. Section 6.7 describes finding different components in networks, like strongest sub-networks or cliques and weakest or articulation points. Section 6.8 provides finding matching solutions for arbitrary patterns based on a path through their all nodes or all links. Section 6.9 is devoted to managing flows in networks including package delivery to the given destinations which may be via arbitrary routes or by using shortest path tree. It also describes effective and extremely compact implementation in SGL of the well known Ford Fulkerson method for finding maximum flow in a graph and its comparison with existing Java solution. Section 6.10 concludes the paper confirming effectiveness of

P. S. Sapaty, *Spatial Networking in the United Physical, Virtual, and Mental World*,
Studies in Systems, Decision and Control 533,
https://doi.org/10.1007/978-3-031-62154-3_6

SGT for basic network operations, and References cite many publication sources analyzed and used in the chapter.

6.2 A Review of Publications on Main Network and Graph Operations

The following sources were analyzed and investigated in detail for drafting the repertoire of main networking operations which may be useful, even critical, for their usage within areas described in Chap. 2.

An introduction to networks is in [74]. A network is simply a collection of connected objects. We refer to the objects as nodes or vertices, and usually draw them as points. We refer to the connections between the nodes as edges, and usually draw them as lines between points. Networks can represent all sorts of systems in the real world. For example, one could describe the Internet as a network where the nodes are computers or other devices and the edges are physical (or wireless, even) connections between the devices. The World Wide Web is a huge network where the pages are nodes and links are the edges. Other examples include social networks of acquaintances or other types of interactions, networks of publications linked by citations, transportation networks, metabolic networks, and communication networks. You can click on the following images for more information about their respective networks.

The degree distribution of a network is in [75]. A network can be an exceedingly complex structure, as the connections among the nodes can exhibit complicated patterns. One challenge in studying complex networks is to develop simplified measures that capture some elements of the structure in an understandable way. One such simplification is to ignore any patterns among different nodes, but just look at each node separately. If one zooms in onto a node and ignores all other nodes, the only thing one can see is how many connections the node has, i.e., the degree of the node.

Random networks are in [76]. When analyzing a network, one approach is look at the network as a single fixed entity. But sometimes, it is useful to think of the edges as random variables. With this random network perspective, a given network is more than a single object. Instead, we can view the random network as a sample from a probability distribution. We can then study the whole probability distribution to gain insight into the network. We can view the probability distribution of random networks as defining a whole ensemble of many different networks, where the probability of any particular network is determined by the probability distribution. Rather than looking at the particular properties of a single network, we can study the properties of the whole ensemble of networks.

Network algorithms are in [77]. Network structures characterize how networks "look"—large or small diameter, number of edges, sparse or dense, degree distributions, clustering, etc. The distance between two vertices is the length of the shortest path connecting them. This assumes the network has only a single connected component. If two vertices are in different components, their distance is infinite. The diameter of a network is the maximum distance between a pair of vertices in the network. Eccentricity as the length of the longest shortest path.

10 Fundamental network algorithms are discussed in [78]. This looks at the algorithms used to perform network calculations. It starts with some simple algorithms for calculating quantities such as degrees, degree distributions, and clustering before moving on to more sophisticated algorithms for shortest paths, betweenness, maximum flows, and other non-local quantities. Exercises are provided at the end of the chapter.

Graph and network algorithms are in [79]. Graphs model the connections in a network and are widely applicable to a variety of physical, biological, and information systems. You can use graphs to model the neurons in a brain, the flight patterns of an airline, and much more. The structure of a graph is comprised of "nodes" and "edges". Each node represents an entity, and each edge represents a connection between two nodes.

Distributed network algorithms are in [80]. Distributed network algorithms play a major role in many networked systems, ranging from computer networks (such as sensor networks, peer-to-peer networks, software-defined networks, datacenter networks, networks on chip) to social or even biological networks. The goal of this course is to introduce the fundamental formal models and methods needed to reason about the correctness and performance of distributed network algorithms.

Data structures and network algorithms are in [81]. There has been an explosive growth in the field of combinatorial algorithms. These algorithms depend not only on results in combinatorics and especially in graph theory, but also on the development of new data structures and new techniques for analyzing algorithms. Data Structures and Network Algorithms attempts to provide the reader with both a practical understanding of the algorithms, described to facilitate their easy implementation, and an appreciation of the depth and beauty of the field of graph algorithms.

What is graph in data structure and types of graph are in [82]. Graphs in data structures are non-linear data structures made up of a finite number of nodes or vertices and the edges that connect them. Graphs in data structures are used to address real-world problems in which it represents the problem area as a network like telephone networks, circuit networks, and social networks. For example, it can represent a single user as nodes or vertices in a telephone network, while the link between them via telephone represents edges.

Graph operations are discussed in [83]. Graphs are often a good representation for problems involving objects and their relationships because there are standard graph operations that can be used to answer useful questions about those relationships. Here we discuss two such operations: depth-first search and breadth-first search, and some of their applications. Both depth-first and breadth-first search are "orderly" ways to traverse the nodes and edges of a graph that are reachable from some starting node. The main difference between depth-first and breadth-first search is the order in which nodes are visited.

Basic graph theory is discussed in [84]. This considers such features as: undirected and directed graphs, neighborhood and degree, density and average degree, paths, loops, paths in directed graphs, length, distance, network structure, connected graphs and connected components, hubs, clusters, network dynamics, random graph algorithm, clusters and homophily, triadic closure algorithm, hubs and cumulative advantage, preferential attachment algorithm, and many others.

Graph data structure is explained in [85]. Graphs are data structures that consist of nodes or vertices linked by edges. Graphs are used to depict relationships and links between diverse parts, making it possible to simulate and evaluate complicated systems more efficiently. Data structures are specialized formats that we use to store, process, and retrieve data according to our needs. There are several data structures that serve different purposes depending on the requirements. Some basic operations in a graph are considered like: add/remove vertex or edge, breadth-first search, and depth-first search. Different types of graphs also discussed, like: finite, infinite, trivial, simple, multi, null, complete, pseudo, regular, bipartite, labelled, direct, connected, disconnected, cyclic, and acyclic. Also representation of graphs by adjacency matrix or list.

Ford-Fulkerson algorithm for maximum flow problem is provided in [86]. In Graph Theory, maximum flow is the maximum amount of flow that can flow from source node to sink node in a given flow network. The Ford-Fulkerson algorithm is used to detect maximum flow from start vertex to sink vertex in a given graph. In this graph, every edge has the capacity. Flow can be defined as any physical/virtual thing that we can transmit between two vertices through an edge in a graph, for example—water, current, data and in this case "trains". Assuming steady conditions (Number of trains on a railway track at any instance of time must not exceeds the track's capacity), if we need to find what is the maximum number of trains (maximum flow) that we can send from City A to city B then the problem is known as the "Maximum Flow Problem".

As the summary on the analyzed various networking sources [1–55, 74–86], we selected and collected the number of networking features particularly suitable for the use of SGT and SGL for their expression, analysis and processing. These features include the following: *a directed network*, *an undirected network*, *different types of nodes and edges*, *different weights of edges and nodes*, *degree of the node*, *diameter of a network*, *eccentricity* as the length of the longest shortest path, *clustering*, *betweenness*, *breadth-first* graph search, *depth-first* graph search, *shortest path* between two single nodes, *shortest path tree* from node, *all paths* between two graph nodes, *maximum flow* in graph, *all cycles* in graph, *small-world networks*, *network partition*, *heaps*, *linking and cutting trees*, *minimum spanning trees*, *network flows*, *matchings*, *connected graphs*, *connected components*, and *hubs*.

A number of these and other issues have been already investigated and implemented using SGT, which can be found in the existing publications [56–73]. But some very basic and fundamental features, including very new networking solutions, are provided and explained in detail in the following sections.

6.3 Basics of Spatial Grasp Model and Technology

Only main SGT ideas are mentioned below, more can be found in the related patent [56], previous books [57–65], latest journal papers [66–73], numerous previous publications on this paradigm, as well as the technology description and explanation provided by Chaps. 3–5.

Within Spatial Grasp Model and Technology (SGT), a high-level operational scenario expressed in recursive Spatial Grasp Language (SGL), starting in any world point *propagates, covers, and matches the distributed environment in parallel wavelike mode*. as symbolically shown in Fig. 6.1, Such propagation can result in returning and analyzing the reached states and data, which may be arbitrarily remote, or used for launching more waves.

SGL allows for organizing *direct space presence and operations* with unlimited powers and parallelism. Its universal recursive organization, with operational scenarios called *grasp*, can be expressed by a single formula:

$$\textbf{grasp} \rightarrow \textbf{constant} \mid \textbf{variable} \mid \textbf{rule}\ (\{\textbf{grasp},\})$$

The *rule* expresses certain action, control, description or context accompanied with operands, which can be any *grasps* too. Rules, starting in certain points, can organize navigation of the world sequentially, in parallel, or any combinations thereof. They can result in the same application points or cause movement to other world points with obtained results left there or returned.

The *SGL interpreter* consists of specialized modules serving SGL scenarios or their parts happened to be inside this interpreter, also organizing exchanges with other interpreters for distributed SGL scenarios. Communicating SGL interpreters can be in arbitrary number of copies, representing *powerful spatial engines operating without central resources or control*, as in Fig. 6.2. They can effectively work with *spatial graph and network data* of any volumes and distributions.

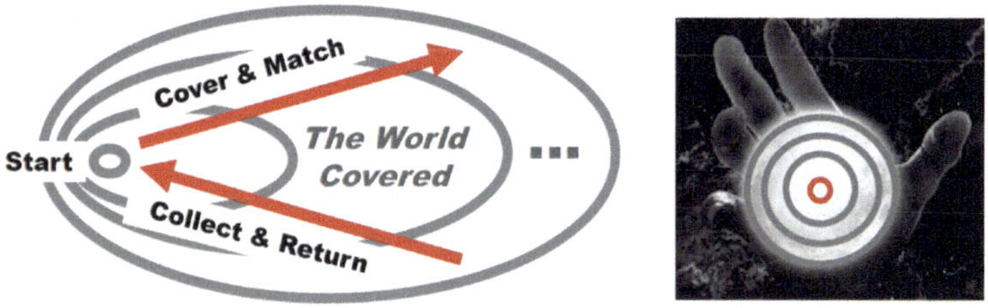

Fig. 6.1 Parallel wavelike world coverage under Spatial grasp model

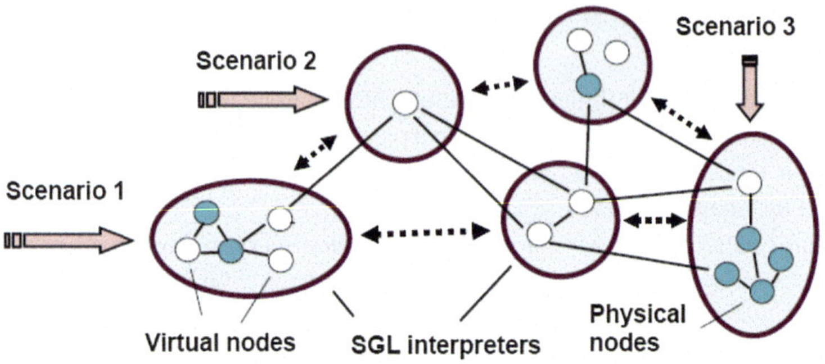

Fig. 6.2 SGL distributed networked interpretation

6.4 Elementary Networking Operations in SGL

- **Creating individual nodes, links**

Create isolated virtual node with some name a (see Fig. 6.3a):

```
create_node(a)
```

The physical location, size, and shape of such node can be added as follows, after hoping into the already created node and staying in it:

```
hop_node(a); WHERE = X_Y, SIZE = big; SHAPE = star.
```

These full details can also be set up during the node creation:

```
create_node(NAME:a, WHERE:X_Y; SIZE:big, SHAPE:star)
```

Adding to the single node a link named r into another node b, to be created too, will be as (see Fig. 6.3b):

```
hop(a); create(link(r), node(b));
```

Explicitly adding the created link details, when staying the node it leads to, after passing the link, will be as follows:

```
create(link(r), node(b));
ORIENT = +, WEIGHT = 10; LENTH = 50
```

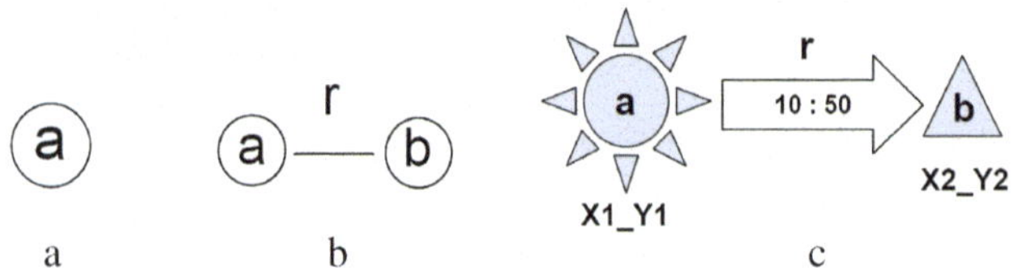

Fig. 6.3 Representation of network nodes and links

Or providing the full link details during its creation from the starting node:

```
hop(a); create(link(LINK:r, ORIENT:+, WEIGHT:10,
                  LENGTH:50), node(b))
```

Full details of both nodes and link connecting them can be set up as follows (see Fig. 6.3c):

```
create_node(NAME:a, WHERE:X1_Y1, SIZE:big,
            SHAPE:star);
create(link(LINK:r, ORIENT:+, WEIGHT:10, LENGTH:50),
       node(NAME:b, WHERE:X2_Y2, SIZE:small,
            SHAPE:triangle))
```

```
SHAPE:triangle)).
```

- **Network creation**

Crating tree-like network, as in Fig. 6.4a:

```
create(node(a);
       ((link(r), node(b)); (link(r), nodes(d, e))),
       ((link(r), node(c)); (link(r), nodes(f, g))))
```

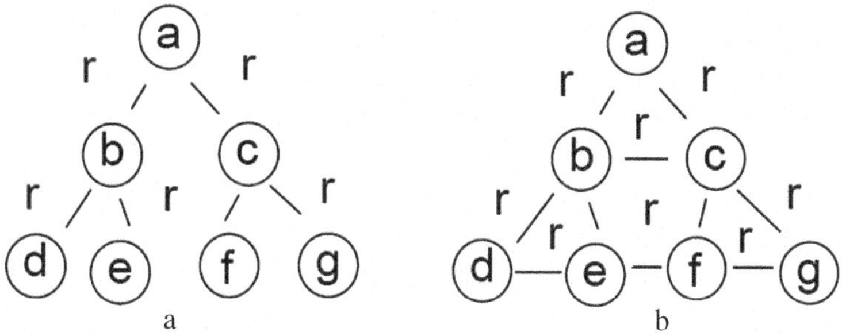

Fig. 6.4 Tree-like and arbitrary network topologies

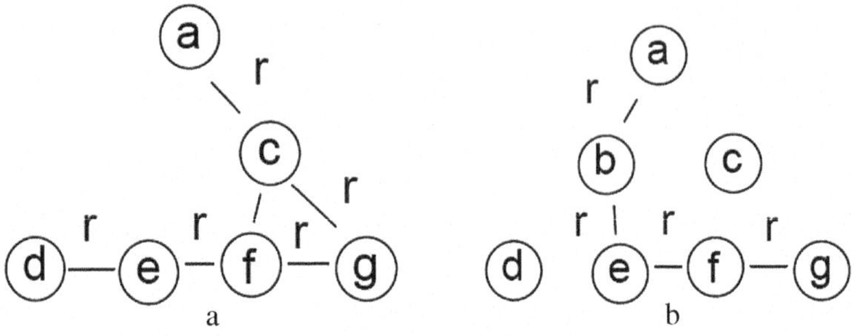

Fig. 6.5 Removing nodes and links from the network

Creating arbitrary network, as in Fig. 6.4b:

```
Top = (a:(b,c), b:(a,c,d,e), c:(a,b,f,g), d:(b,e),
      e:(b,d,f), f:(e,c,g), g:(c,f));
align(split(Top); frontal(Next) = VAL[2];
      create(VAL[1]));
hop(Next); NAME > BEFORE; linkup(r, BEFORE)
```

- **Nodes and links removal**

Remove node b of Fig. 6.4b, see the result in Fig. 6.5a:

```
remove(hop_node(b))
```

Remove all links associated with nodes c and d of Fig. 6.4b, with the result in Fig. 6.5b.

```
hop_nodes(c, d); remove(all_links)
```

6.5 Finding Paths in Networks

- **Finding and collecting any path between certain nodes**

Having created a network infrastructure, like the one as in Fig. 6.4b, we may, starting in some node like d, reach any other node like c, by the following SGL scenario navigating the network in wavelike mode, i.e. stepwise, in parallel, and without cycles (see Fig. 6.6 with different solutions for this tas).

```
hopfirst(d);
repeat(hopfirst(all_links); if(NAME == c, done))
```

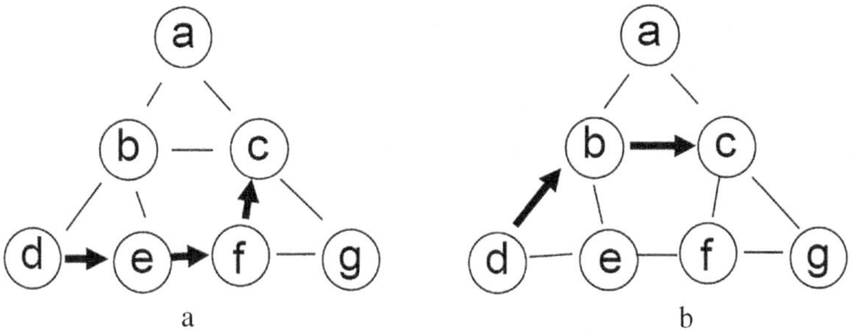

Fig. 6.6 Reaching a node from another node

We may additionally decide to collect the passed path and organize its output at the destination node c, as follows.

```
hopfirst(d); frontal(Path = NAME);
repeat(hopfirst(all_links); Path &&= NAME;
        if(NAME == c, done_output(Path)))
```

This will give (d,e,f,c) for Fig. 6.6a, and (d,b,c) for Fig. 6.6b. Such paths can also be issued in the starting node d by the modified scenario:

```
hopfirst(d); frontal(Path = NAME);
output_repeat(hopfirst(all_links); Path &&= NAME;
              if(NAME == c, blind(Path)))
```

- **Finding and collecting all paths between certain nodes in the network**

It is easy to find any paths in networks in SGL. For example, all simple paths from node d to node c (i.e. with non repeating nodes) can be easily found from Fig. 6.4b by:

```
frontal(Path); hop(d);
repeat(append(Path, NAME);
       if(NAME == c, done_output(Path));
       hop(all_links); notbelong(NAME, Path))
```

If we apply this scenario to the network of Fig. 6.4b, will receive the following paths:

```
(d,b,c), (d,b,a,c), (d,e,b,a,c), (d,e,b,c),
(d,e,f,c), (d,e,f,g,c)
```

6.6 Spanning Trees

- **Breadth-first spanning tree (BST) from a node to all other nodes**

In one of the previous examples, we have found a possible path from a node to some particular node, which was then issued outside the network within the network asynchronous wavelike navigation. In a similar network navigation, we may create a Breadth-first Spanning Tree (BST) starting in the same start node and covering the whole network, thus explicitly showing possible paths to all other nodes. This can be easily done with registering this BST directly in the distributed network structure, by remembering predecessor node names in a special variable `Up` associated with each node, as follows, see also Fig. 6.7a, b for possible BST solutions:

```
nodal(Up); hopfirst(d);
repeat(hopfirst(links_all)); Up = BACK)
```

Such a tree can directly guide movement from any node, like `c`, to the starting node `d`:

```
hop(c); repeat(hop(Up))
```

We may also collect this passed path as follows.

```
hop(c); Path = NAME;
repeat(hop(Up); Path = NAME && Path); output(Path)
```

Results: `(d,e,f,c)` for Fig. 6.7a and `(d,b,c)` for Fig. 6.7b.

- **Depth-first spanning tree (DST) from a node to all other nodes**

See Fig. 6.8a, b for possible DST solutions, with starting from node d:

```
DST = {sequence(hopfirst(links_all); Up = BACK;
        run(DST))};
hopfirst(d); run(DST)
```

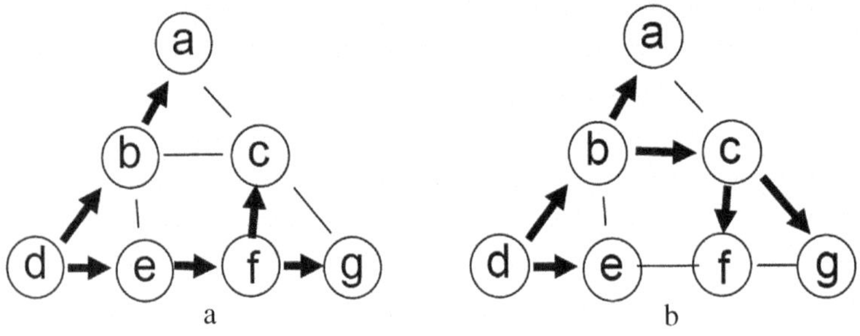

Fig. 6.7 Creating any spanning tree from start `d` to all other nodes

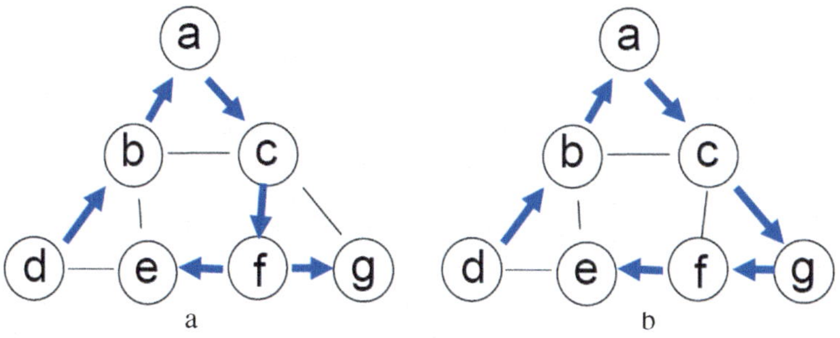

Fig. 6.8 Possible depth-first spanning trees from start d to all other nodes

Collecting path from start d to any other node, say, e:

```
hop(e); Path = NAME;
repeat(hop(Up); Path = NAME && Path); output(Path)
```

Result: (d,b,a,c,f,e) for Fig. 6.8a, or (d,b,a,c,g,f,e) for Fig. 6.8b.

- **Shortest path tree (SPT) from a node to all other nodes**

With some modification, we may create and register instead of any spanning tree, a Shortest Path Tree (SPT) from a node to all other nodes (which, in general, may have more than a single solution, as shown in Fig. 6.9a, b).

The following scenario can accomplish this with registering the resultant SPT in the network structure similarly to the spanning tree scenario before. The main difference will be with re-registering the SPT predecessor nodes in variables Up if shorter solutions from the start to these nodes appear available.

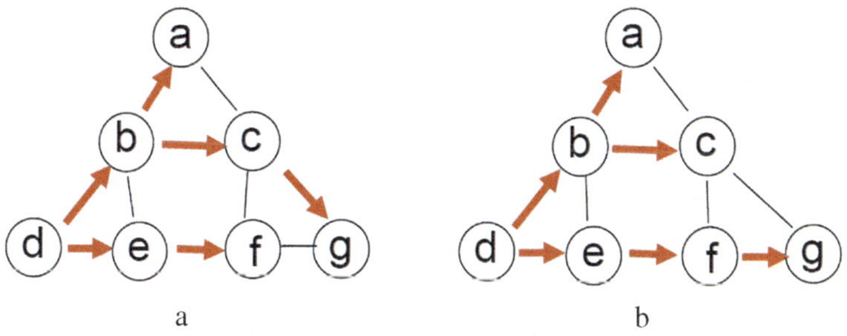

Fig. 6.9 Creating shortest path tree from start d to all other nodes

```
nodal(Dist, Up); hop(d); Dist = 0; frontal(Far);
repeat(
  hop(links_all); Far += 1;
  or(Dist == nil, Dist > Far);
  Dist = Far; Up = BACK)
```

SPT will be embedded into the network structure with `Dist` and `Up` in all nodes. Starting in any node, say `c`, and following `Up` you can easily receive SP to it from the head of the SPT, similar to BST above:

```
hop(c); Spath = NAME;
repeat(hop(Up); Path = NAME && Path); output(Spath)
```

Result of SP from `d` to `c`: `(d,b,c)` for both SPT in Fig. 6.9.

6.7 Finding Different Components in Networks

- **Finding strongest sub-networks, or cliques**

We present here a universal solution in SGL for finding all cliques in a network (i.e. maximum possible full subgraphs), assuming their number of nodes should correspond to some threshold. The following scenario is finding all cliques in parallel in the network of Fig. 6.10 with the number of nodes not less than three (only one clique, with four nodes, is highlighted in the figure).

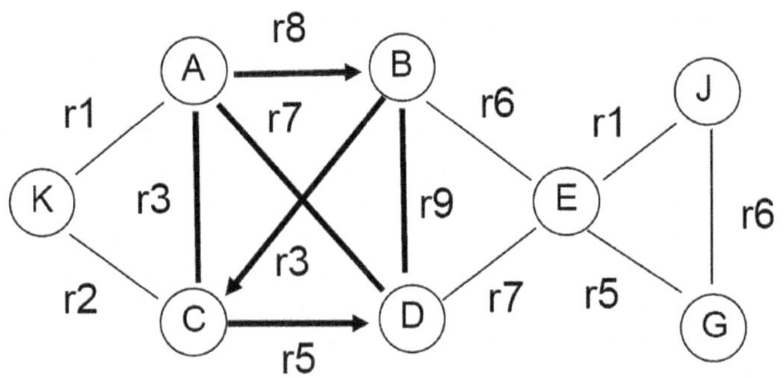

Fig. 6.10 Finding strongest parts, or cliques, in the network

```
hop_nodes(all); frontal(Clique) = NAME;
repeat(
 hop_links(all); not_belong(NAME, Clique);
 yes(and_parallel(hop(links_any, nodes(Clique))));
 if(PREDECESSOR > NAME,
    append(Clique, NAME), blind));
count(Clique) >= 3; output(Clique)
```

This scenario, starting in all nodes and following their links to other nodes in parallel, is collecting node names in new individual hops which have links with all previously collected nodes unless such nodes cannot be found, declaring the collected set of node names in the frontal variable `Clique` as a new clique. As the same clique can be collected in such a way when started from all nodes of the same clique, the duplicates are blocked by allowing inclusion of any new node into the clique's list only if the value of its name is lower than of the previous one. The obtained full results for Fig. 6.10 will be as:

```
(A,B,C,D), (A,C,K), (B,D,E), (E,G,J)
```

If names of links in cliques are of interest too, the modified scenario will look like:

```
 hop_links(all); not_belong(NAME, Clique);
 yes(and_parallel(hop(links(any), nodes(Clique))));
 if(PREDECESSOR > NAME,
    (append(Clique, NAME);
     append(
       Links,(hop(links_any, nodes(Clique)); LINK))),
    blind));
count(Clique) >= 3; output(unit(Clique, Links))
```

The obtained results will be as follows:

```
(A,B,C,D,r8,r3,r3,r5,r7,r9), (A,C,K,r3,r2,r1),
(B,D,E,r9,r7,r6), (E,G,J,r5,r6,r1)
```

- **Discovering weakest, or articulation points**

Weakest or articulation points when removed split the network into disjoint parts, like node `E` in Fig. 6.11.

Next is parallel and fully distributed solution for finding all articulation points in the network, by which each network node, first selecting one neighbor randomly,

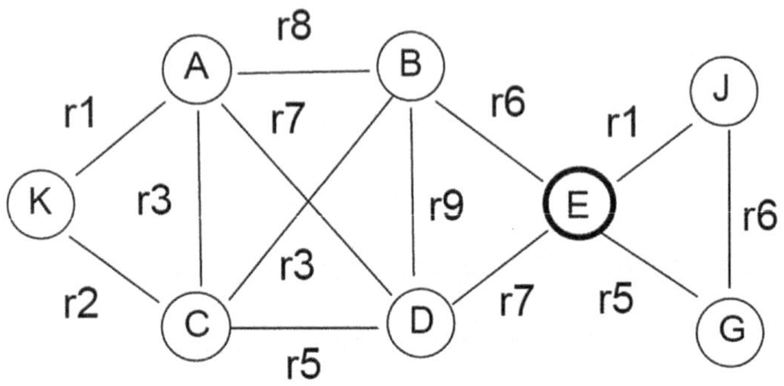

Fig. 6.11 Weakest or articulation points of the network

tries to navigate and mark the whole network from it while excluding itself from this process. After termination, if the node discovers still unmarked neighbors, it declares itself articulation point and outputs its name, as follows:

```
hop_nodes(all); IDENTITY = NAME;
hopfirst_node(current);
stay(hopfirst_random(links_all);
     repeat(hopfirst(links_all)));
if(hopfirst(links_all), output(NAME))
```

The answer will be E, as the only point of such type in the network of Fig. 6.11.

6.8 Pattern Matching

- **Arbitrary network matching patterns**

In distributed networks, like shown in Fig. 6.12a, any graph structures can be found with the use of SGT, for example, as the one shown in Fig. 6.12b. Applied in parallel from all network nodes, it can then self-evolve from them in parallel matching mode with finding meanings of variables N1 to N4. There may be different possibilities of how to represent arbitrary search patterns as matching templates in SGL.

- **Template representation based on a path through all nodes**

Among simplest matching template representations may be the one based on a path through all pattern's nodes, as shown in Fig. 6.13b (for more complex cases such a path may have to include some nodes more than once). The SGL self-matching scenario based on the oriented path of Fig. 6.13b will be as follows.

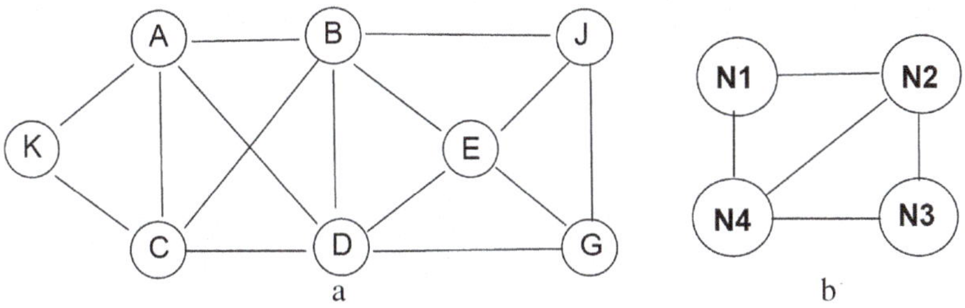

Fig. 6.12 A network and its search pattern

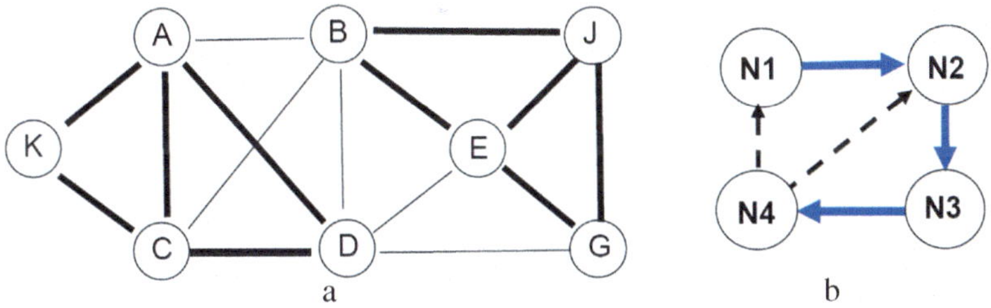

Fig. 6.13 Matching with template based on a path trough all nodes

```
hop_nodes(all); frontal(Nodes) = NAME;
three_repeat(hop_link(all); notbelong(NAME, Nodes);
                Nodes && = NAME);
true(andparallel(link(any), Nodes[1,2]));
```

The nodes of found substructures corresponding to variables `N1` to `N4` will be as follows (also taking into account that the pattern may start from different nodes of the same network structures, with only two solutions highlighted in Fig. 6.13a):

```
(N1,N2,N3,N4) ➔ (K,A,D,C), (B,J,G,E), (A,B,E,D), ...
```

The output of contents of variables found will be accomplished in nodes corresponding to variable `N4`, which will be `C`, `E`, `D` and others, in another matching.

- **Template representation based on a path through all links**

Another simple template representation in SGL, fully universal too, can be based on a path through all template links, as in Fig. 6.14b (with numbers of variables on links corresponding to the order of their passing), where in general some links may have to be passed more than once. Such template representation is especially convenient for the joint collection of the matched node and link names, as shown in Fig. 6.14a, b, with variables `Ni` for nodes, and `Li` for links. The corresponding SGL scenario

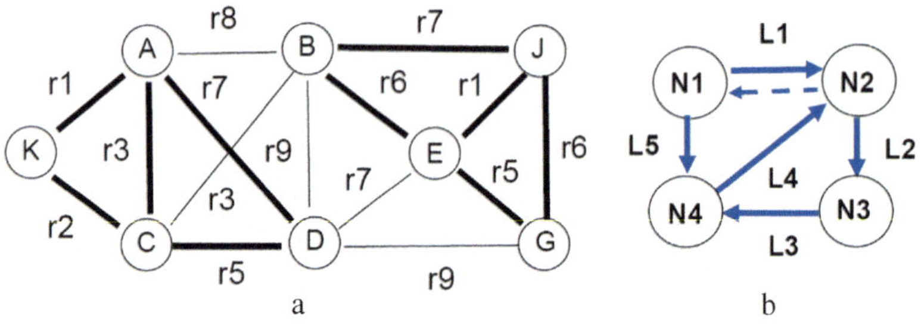

Fig. 6.14 Matching with template based on a path trough all links

for finding and issuing values of all these variables in the successful matches will be as follows.

```
hop_nodes(all); frontal(Nodes = NAME, Links);
three_repeat(hop_link(all); notbelong(NAME, Nodes);
                 Links &&= LINK; Nodes &&= NAME);
hop(link(all), Nodes[2]); Links &&= LINK;
hop(Links[1], Nodes[1]);
hop(link(any), Nodes[4]); Links &&= LINK;
output(Nodes, ":", Links)
```

Some of the solutions with variables at both nodes and links will be as follows (issued in nodes corresponding to variable N4, same as in previous example of Fig. 6.13):

```
(N1,N2,N3,N4):(L1,L2,L3,L4,L5) ➔
(K,A,D,C):(r1,r7,r5,r3,r2),
(B,J,G,E):(r7,r6,r5,r1,r6), ...
```

6.9 Deliveries and Flows in Networks

- **Package delivery to the given destinations**

Reaching from `Start` node the set of particular `Fin` nodes (using any simple casual paths to them, i.e. without loops) and delivering a sort of `Pack` to them (like attaching it to their `CONTENT`), may be expressed as follows, see also Fig. 6.15.

Fig. 6.15 Possible delivery paths from `Start` to certain destinations

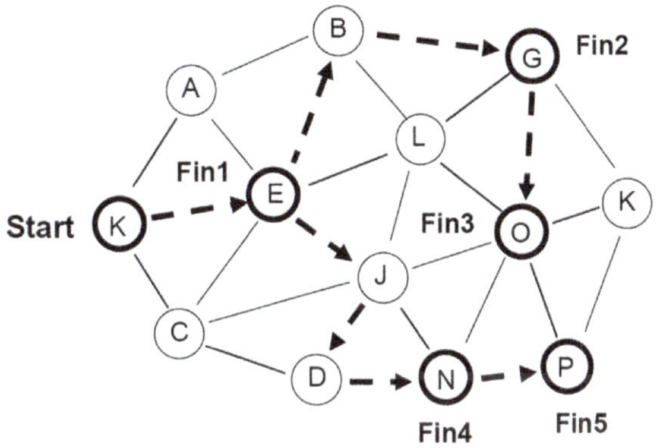


```
frontal(Pack = …, Fin = (E, G, N, O, P));
nodal(Start = K);
hopfirst(Start);
repeat(if(belong(NAME, Fin), CONTENT &&= Pack);
       hopfirst(links_any))
```

With not optimal casual delivery paths of this case, there may be inefficiency and redundancy of the network usage, also delays with deliveries.

- **Optimal package delivery using SPT**

We may first find shortest path tree of this network, as in Fig. 6.16a, originating from `Start`, and then use only paths in it leading to the `Fin` nodes (as in Fig. 6.16b), altogether forming the best delivery infrastructure from `Start` to `Fin` (some of its nodes may happen to be on the same shortest paths from `Start`). The related delivery scenario may be as follows.

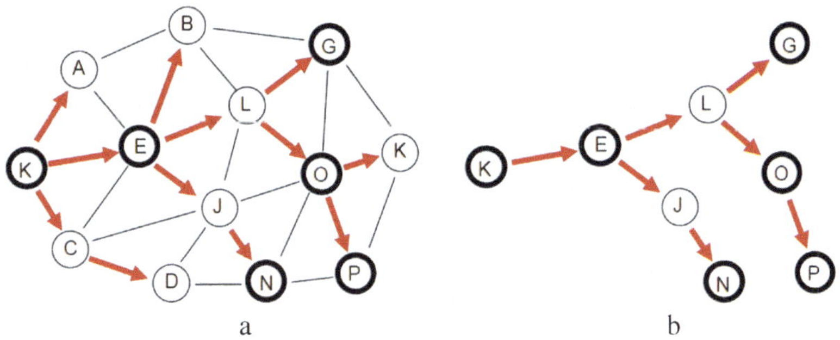

Fig. 6.16 Forming shortest path delivery infrastructure to certain destinations

```
nodal(Dist, Up, Up2, Fin = (E,G,N,O,P));
frontal(Far); nodal(Start) = K;
hop(Start);
sequence(
 repeat(hop(links(all)); Far += 1;
        or(Dist == nil, Dist > Far);
        Dist = Far; Up = BACK),
 (hop(Fin); repeat(Up2 = Up; hop(Up))),
 (frontal(Pack = delivery);
  repeat(hop(all_links); Up2 == BACK;
         if(belong(NAME, Fin),
```

- **Maximum flow in a graph**

In graph theory, maximum flow is the maximum amount that can flow from source node to sink node in a given flow network. The Ford-Fulkerson algorithm [86] is used to detect maximum flow from start vertex to sink vertex in a given graph, where every edge has certain capacity, as in Fig. 6.17 (with initial flow shown as zeros). In simple terms, Ford-Fulkerson Algorithm is: As long as there is a path from source A to sink F with available capacity on all its edges, we may send the possible flow from that path and find another path, and so on. Flow in the network has the following restrictions: Input flow must match to output flow for each node in the graph, except the source and sink node; Flow out from source node must match with the flow in to sink node. Flow from each edge should not exceed the capacity of that node. The following SGL solution is based on the algorithm described in [86].

We start with creation of the network of Fig. 6.17:

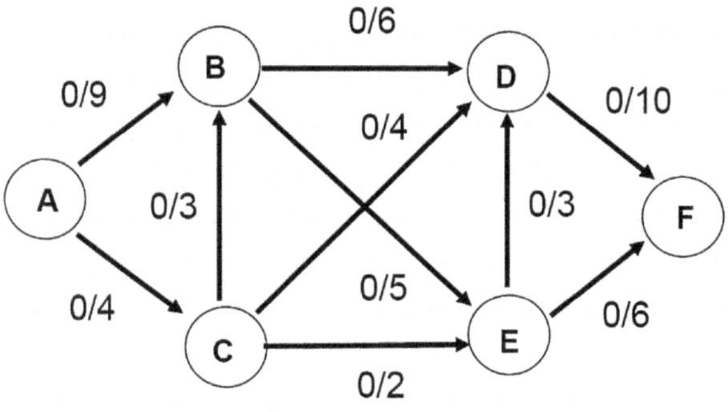

Fig. 6.17 Initial network state for finding maximum flow

```
Top =(A:(9:B,4:C),B:(6:D,5:E),C:(3:B,4:D,2:E),
     D:(10:F),E:(3:D,6:F),F);
align(split(Top); frontal(Next) = VAL[2];
      create(VAL[1]));
split(Next); linkup(+VAL[1], VAL[2]); THRU = 0
```

The following scenario is incrementally finding maximum flow in the network of Fig. 6.17, each time selecting a single path from starting node A to final one F (from all possible paths traversed in parallel) through which the flow can still be increased, like the one in Fig. 6.18a, with amendment the flow through this path, until this remains possible. One of possible final solutions is shown in Fig. 6.18b.

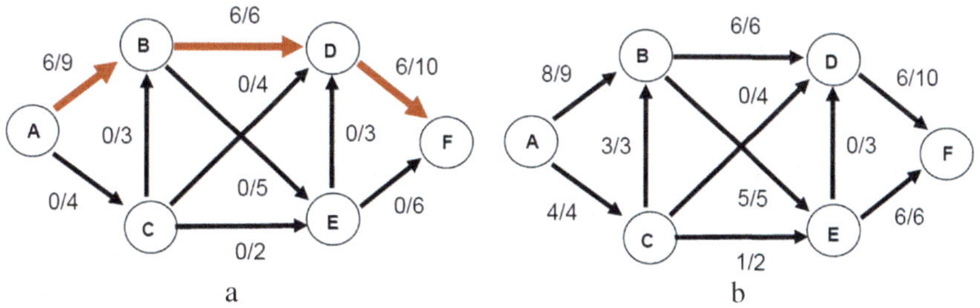

Fig. 6.18 Intermediate and final maximum flow processing

```
frontal(Flow = infin, Path);
Proc =
 {repeat(
  hop(+all_links); notbelong(NAME, Path);
  Flow = min(LINK - THRU, Flow);
  Flow != 0; append(NAME, Path);
  if(NAME == F,
     done_global(
       Result += Flow;
       repeat(move(-link, withdraw(Path));
              THRU += Flow))))};
hop(A); repeat(if(run(Proc), stay, done(hop(F);
               output(Result))))
```

This maximum flow implementation in SGL is much clearer and more compact than in Java (its solution copied in Fig. 6.19) or different versions of C [86], as fundamentally based on *direct spatial presence, thinking, and pattern matching* provided by Spatial Grasp Model and Technology, see for example [56, 59, 60, 62, 71].

```
class MaxFlow{
   // Initializing V=6 because
   // number of vertices is 6.
   static final int V=6;
   public boolean bfs(int residual_g[][],int s,int t,int par
      // visited array will keep track
      // of all the visited vertices.
      boolean visited[]=new int[V];

      // Queue to keep vertices in Queue
      // for doing BFS.
      Queue<Integer> q=new LinkedList<Integer>();
      // Add source in q and mark it visited.
      q.add(s);
      visited[s]=true;
      // Make parent of source -1.
      parent[s]=-1;

      // Standard BFS loop
      while(!q.isEmpty()){
         int u=q.poll();
         for(int v=0;v<V;v++){
            if(!visited[v]&&residual_g[u][v]>0){

               visited[v]=true;
               q.add(v);
               parent[v]=u;

               // If we found a connection to sink
               // from any of the nodes. We will return
               // true i.e. augmenting path exists.
               if(v==t)
                  return true;
            }
         }
      }
      // If we can't reach to sink node then
      // return false which will indicate
      // that there is no augmenting path left.
      return false;
   }
   public int Ford_Fulkerson(int graph[][],int s,int t){

      // Creating a residual graph with residual
      // capacity as the capacity of the original graph.
      int residual_g[][]=new int[V][V];

      // residual_g[i][j] indicates residual capacity
      // of edge from i to j. If residual_g[i][j]==0
      // it means no edge exists.
      for(int i=0;i<V;i++)
         for(int j=0;j<V;j++)
            residual_g[i][j]=g[i][j];

      // Parent array will store the augmenting
      // path found during the BFS.
      int parent[]=new int[V];

      // Initializing max flow as 0 i.e. there
      // is no flow initially.
      int max_flow=0;

      // Iterating in breadth-first manner till there
      // is augmenting path from source to sink.
      while(bfs(residual_g,s,t))
      {
         // Initializing flow of the current path
         // with a very big value.
         int flow=Integer.MAX_VALUE;

         // Finding minimum residual capacity of the
         // edges in augmenting path i.e. maximum
         // flow through that path.
         for(int v=t;v!=s;v=parent[v])
         {
            int u=parent[v];
            flow=Math.min(flow, residual_g[u][v]);
         }
         // Adding maximum flow of this path
         // to the overall flow.
         max_flow+=flow;

         // Updating residual capacities of edges
         // and the back edges along the path.
         for(int v=t;v!=s;v=parent[v])
         {
            int u=parent[v];
            // Decreasing capacity of the forward edge
            residual_g[u][v]-=flow;
            // Increasing capacity of the backward edg
            residual_g[v][u]+=flow;
         }
      }

      // Returning our answer
      // i.e. Maximum flow through
      // the network.
      return max_flow;
   }
```

Fig. 6.19 Java version of Ford-Fulkerson method

6.10 Conclusions

The results of this chapter confirm availability and efficiency of using SGT and SGL for investigating, expressing, and implementing most basic network and graph creation and management issues. SGL allows us to provide detailed, very clear and extremely compact solutions for the network problems which can operate on arbitrary large and complex networks, and in highly parallel and fully distributed mode. *The solutions provided and discussed in this paper were obtained in a direct spatial thinking, spatial pattern recognition, and spatial pattern matching mode, rather than under traditional logic-based and algorithmic philosophy and culture.* Being highly parallel and fully distributed, they may challenge existing opinions that parallel and distributed algorithms are usually much more complex than traditional sequential ones for solving same problems, just proving the opposite, and especially for the network-related problems.

References

1. Transport networks, Inspire Registry (2015). https://inspire.ec.europa.eu/theme/tn
2. Transportation networks, Travel Forecasting Resource. https://tfresource.org/topics/Transportation_networks.html
3. Durkin, J., Goodman, J., et al.: Building Multiservice Transport Networks (Networking Technology), Cisco Press; 1st end (2012)
4. Murakami, M., Li, H., Ryoo, J-d.: Packet Transport Networks: Overview and Future Direction, 2 nd APT/ITU Conformance and Interoperability Workshop (C&I-2) (2014). https://www.itu.int/en/ITU-T/C-I/interop/Documents/20140826/CI-2_INP-09_Packet_Transport_Networks_Overview_and_Future_Direction.pdf
5. Air Transport Network. https://en.wikipedia.org/wiki/Air_Transport_Network
6. Qureshi, K.N., Ul Islam, F., et. al.: Improved road segment-based geographical routing protocol for vehicular Ad-hoc networks. Electronics **9**(8), 1248 (2020). https://doi.org/10.3390/electronics9081248, https://www.mdpi.com/2079-9292/9/8/1248
7. Sujan: What is communication network? Definition, Importance, and 4 Types (Explained), June 24, 2023. https://tyonote.com/communication_network/
8. Fitzek, F.H.P., Granelli, F., et al.: Computing in Communication Networks: From Theory to Practice, Academic Press; 1st edn. (2020)
9. Srikant, R., Ying, L.: Communication Networks: An Optimization, Control, and Stochastic Networks Perspective. Cambridge University Press, 1st edn. (2013)
10. Raghunandan, K.: Introduction to Wireless Communications and Networks: A Practical Perspective (Textbooks in Telecommunication Engineering), 1st edn. Springer (2022). April 2, 2023
11. Lou, Y., Ahmed, N.: Underwater Communications and Networks (Textbooks in Telecommunication Engineering). Springer (2021)
12. Kaufman, C., Perlman, R., et al.: Network Security: Private Communication in a Public World (Prentice Hall Series in Computer Networking and Distributed Systems), 3rd edn. Addison-Wesley Professional (2022)
13. Wu, Y., Singh, S. et al.: 6G Mobile Wireless Networks (Computer Communications and Networks), 1st edn. Springer (2021). August 26, 2022
14. Social network. https://en.wikipedia.org/wiki/Social_network

15. Kenton, W.: What is social networking? Investopedia, Updated May 24, 2023. https://www.investopedia.com/terms/s/social-networking.asp#:~:text=Social%20networking%20refers%20to%20using,)%2C%20Instagram%2C%20and%20Pinterest
16. Nanawati, S.: Social Network Analytics, Published in Analytics Vidhya. Aug 1, 2019. https://medium.com/analytics-vidhya/social-network-analytics-f082f4e21b16
17. Singh, C.: 20+ Most Popular Social Networking Sites in 2024, December 17, 2023. https://www.socialpilot.co/social-networking-sites
18. Jackson, M.O.: Social and Economic Networks. Princeton University Press; Illustrated edition (2010)
19. Harrison, T.: BattleNetworks and the Future Force, Center for Strategic and International Studies (2021). https://www.csis.org/analysis/battle-networks-and-future-force
20. Stillion, J., Clark, B.: What It Takes To Win: Succeeding In 21st Century BattleNetwork Conmpetitions, Center for Strategic and Budgetary Assessments (CSBA) (2015). https://csbaonline.org/uploads/documents/What-it-Takes-to-Win.pdf
21. Hadley, G.: DAF Outlines a New 'Battle Network' as Its Contribution to JADC2, Air & Space Forces Magazine, March 23, 2023. https://www.airandspaceforces.com/daf-battle-network-contribution-jadc2/
22. The Challenges of Military Networks and Technological Developments, Elbit Systems. https://elbitsystems.com/e-lynx_multi_channel/
23. Themnér, A., Karlén, N. (2020) Building a safety net: explaining the strength of ex-military networks. Secur. Stud. **29**(2), 268–300. https://doi.org/10.1080/09636412.2020.1722851
24. What is an Economic Network? Advantages and disadvantages of economic networks, Vietnambiz (2019). https://vietnambiz.vn/mang-luoi-kinh-te-economic-network-la-gi-uu-nhuoc-diem-cua-mang-luoi-kinh-te-20191011113428212.htm
25. Halton, C.: Economic Network: What it is, Pros and Cons, Examples, Updated October 24, 2022. https://www.investopedia.com/terms/e/economic-network.asp#:~:text=Key%20Takeaways-,An%20economic%20network%20is%20a%20combination%20of%20individuals%2C%20groups%2C%20or,companies%20or%20partnerships%20between%20corporations.
26. Todo, Y., Matous, P.: Economic networks for more innovative and resilient economies, 26 Jun 2015. https://cepr.org/voxeu/columns/economic-networks-more-innovative-and-resilient-economies
27. Emmert-Streib, F., Tripathi, S., et al.: Understanding the world economy in terms of networks: a survey of data-based network science approaches on economic networks. Front. Appl. Math. Stat. 28 August 2018, Sec. Math. Finance **4** (2018). | https://doi.org/10.3389/fams.2018.00037, https://doi.org/10.3389/fams.2018.00037/full
28. Wixted, B.: Innovation System Frontiers: Cluster Networks and Global Value (Advances in Spatial Science). Springer (2009)
29. Manyika, J.: What the economy looks like so far for individuals in the 21st century. Feb 6, 2020. https://www.linkedin.com/pulse/what-economy-looks-like-so-far-individuals-21st-century-james-manyika
30. Hayes, A.: Distribution network: definition, how it works, and examples, investopedia, Updated February 22, 2021. https://www.investopedia.com/terms/d/distribution-network.asp#:~:text=In%20a%20supply%20chain%2C%20a,or%20through%20a%20retail%20network.
31. Distribution Network. The flow of goods from a producer or supplier to an end consumer. Written by CFI Team. https://corporatefinanceinstitute.com/resources/valuation/distribution-network/
32. What Are Types Of Distribution Networks In Supply Chain Management? oboloo FAQ's. https://oboloo.com/blog/what-are-types-of-distribution-networks-in-supply-chain-management/
33. Distribution Network—What Is It And What Are The Benefits? https://www.salesbabu.com/blog/what-is-a-distribution-network-and-its-benefits/
34. Wu, Q., Shen, F., et al.: Optimal Operation of Active Distribution Networks: Congestion Management. Academic Press, Voltage Control and Service Restoration (2023)

35. Watts, S.: What is a virtual network? BMC, June 16, 2020. https://www.bmc.com/blogs/virtual-network/
36. What is Virtual Networking? VMware by Broadcom. https://www.vmware.com/topics/glossary/content/virtual-networking.html
37. Network virtualization. https://en.wikipedia.org/wiki/Network_virtualization.
38. Carlos, O., Duarte, M.B., Pujolle, G.: Virtual Networks: Pluralistic Approach for the Next Generation of Internet (Networks and Telecommunications), 1st edn. Wiley-ISTE (2013)
39. Fundamentals of Virtual Networking. https://www.geeksforgeeks.org/fundamentals-of-virtual-networking/
40. Neural network. https://en.wikipedia.org/wiki/Neural_network
41. Anand, V.: Introduction to Neural Network in Machine Learning, Updated On August 10th, 2023. https://www.analyticsvidhya.com/blog/2022/01/introduction-to-neural-networks/
42. Ashtari, H.: What is a neural network? Definition, Working, Types, and Applications in 2022. August 3, 2022. https://www.spiceworks.com/tech/artificial-intelligence/articles/what-is-a-neural-network/
43. Artificial Neural Network. NVIDIA Developer. https://developer.nvidia.com/discover/artificial-neural-network
44. Wu, L., Cui, P., et al.: Graph Neural Networks: Foundations, Frontiers, and Applications, 1st edn. Springer (2022)
45. Borsboom, D., Deserno, M.K., Rhemtulla, M. et al.: Network analysis of multivariate data in psychological science. Nat. Rev. Methods Primers **1**, 58 (2021). https://doi.org/10.1038/s43586-021-00055-w, https://www.nature.com/articles/s43586-021-00055-w
46. Jones, P.J., Mair, P., McNally, R.J.: Visualizing psychological networks: a tutorial in R. Front. Psychol. Sec. Quant. Psychol. Meas. **9** (2018). | https://doi.org/10.3389/fpsyg.2018.01742, https://doi.org/10.3389/fpsyg.2018.01742/full
47. Fonseca-Pedrero, E.: Network analysis in psychology. Psychol. Pap. **39**(1), 1–12 (2018). https://doi.org/10.23923/pap.psicol2018.2852, https://www.papelesdelpsicologo.es/pdf/2852.pdf
48. Bringmann, L.F.: Person-specific networks in psychopathology: past, present, and future. Curr. Opin. Psychol. **41**, 59–64 (2021). https://www.sciencedirect.com/science/article/pii/S2352250X21000300?via%3Dihub
49. Yuqing, C., Shuyang, D., et al.: Network analysis and its applications in psychology. Adv. Psychol. Sci. **28**(1), 178–190 (2020). https://doi.org/10.3724/SP.J.1042.2020.00178
50. Campana, P.: Explaining criminal networks: strategies and potential pitfalls. Methodol. Innov. **9**, 1–10 (2016). https://doi.org/10.1177/2059799115622748
51. Bichler, G.: Understanding Criminal Networks: A Research Guide; 1st edn. University of California Press (2019)
52. Morselli, C. (2013) Crime and Networks (Criminology and Justice Studies), Routledge
53. New forms of organized crime: networked structure. UNODS, Unite Nations Office on Drugs and Crime. https://www.unodc.org/e4j/en/organized-crime/module-7/key-issues/networked-structure.html
54. Hufnagel, S., Moiseienko, A.: Criminal Networks and Law Enforcement: Global Perspectives On Illegal Enterprise (Transnational Criminal Justice), 1st edn. Routledge (2019)
55. Detecting and Disrupting Transnational Criminal Organizations: Analytics for Interdependent Smuggling and Money-Laundering Networks, Criminal Investigation and Network Analysis, DHS Centers of Excellence. https://cina.gmu.edu/projects/detecting-and-disrupting-transnational-criminal-organizations-analytics-for-interdependent-smuggling-and-money-laundering-networks/
56. Sapaty, P.S.: A distributed processing system. European Patent N 0389655, Publ. 10.11.93, European Patent Office
57. Sapaty, P.S.: Mobile Processing in Distributed and Open Environments. Wiley, New York (1999)
58. Sapaty, P.S.: Ruling Distributed Dynamic Worlds. Wiley, New York (2005)
59. Sapaty, P.S.: Managing Distributed Dynamic Systems with Spatial Grasp Technology. Springer (2017)

60. Sapaty, P.S.: Holistic Analysis and Management of Distributed Social Systems. Springer (2018)
61. Sapaty, P.S.: Complexity in International Security: A Holistic Spatial Approach. Emerald Publishing (2019)
62. Sapaty, P.S.: Symbiosis of Real and Simulated Worlds Under Spatial Grasp Technology. Springer (2021)
63. Sapaty, P.S.: Spatial Grasp as a Model for Space-Based Control and Management Systems. CRC Press (2022)
64. Sapaty, P.S.: The Spatial Grasp Model: Applications and Investigations of Distributed Dynamic Worlds. Emerald Publishing (2023)
65. Sapaty, P.S.: Providing Integrity, Awareness, and Consciousness in Distributed Dynamic Systems, CRC Press (2025)
66. Sapaty, P.S.: Spatial management of air and missile defence operations. Math. Mach. Syst. 1 (2023). http://www.immsp.kiev.ua/publications/articles/2023/2023_1/01_23_Sapaty.pdf
67. Sapaty, P.S.: Providing Distributed System Integrity Under Spatial Grasp Technology. Math. Mach. Syst. 2 (2023). http://www.immsp.kiev.ua/publications/articles/2023/2023_2/02_23_Sapaty.pdf
68. Sapaty, P.S.: Providing global awareness in distributed dynamic systems. Int. Relat. Dipl. **11**(2), 87–100 (2023). https://doi.org/10.17265/2328-2134/2023.02.002. https://www.davidpublisher.com/Public/uploads/Contribute/6486c3d05a6cc.pdf
69. Sapaty, P.S.: Simulating distributed consciousness with spatial grasp model. Math. Mach. Syst. 3 (2023)
70. Sapaty, P.S.: Network centrality operations under spatial grasp technology. J. Adv. Artif. Intell. Mach. Learn. **1**(1), 1–11 (2023). https://www.scivisionpub.com/journals/articleinpress-journal-of-advances-in-artificial-intelligence-and-machine-learning
71. Sapaty, P.S.: Managing distributed systems with spatial grasp patterns. Math. Mach. Syst. 4 (2023)
72. Sapaty, P.S.: Spatial networks as models for organoid cultures and brain research. Math. Mach. Syst. 1 (2024)
73. Sapaty, P.S.: Basic network operations under spatial grasp model. Sci. Set. J. Phys. **3**(1), 01–16 (2024)
74. An Introduction to Networks. https://mathinsight.org/network_introduction
75. The Degree Distribution of a Network. https://mathinsight.org/degree_distribution
76. Random Networks. https://mathinsight.org/random_network
77. Network Algorithms. https://ics.uci.edu/~goodrich/teach/cs165/notes/NetworkAlgs.pdf
78. Newman, M.E.J.: 10 fundamental network algorithms: a discussion of some of the most important and fundamental algorithms for performing network calculations on a computer (2010). https://doi.org/10.1093/acprof:oso/9780199206650.003.0010
79. Graph and Network Algorithms. https://ch.mathworks.com/help/matlab/graph-and-network-algorithms.html
80. Schmid, S., Mandal, P.S.: Distributed Network Algorithms (Lecture Notes for GIAN Course) (2016). https://schmiste.github.io/GIAN-Lecture-Notes-NetAlg.pdf
81. Tarjan, R.E.: Data Structures and Network Algorithms, CBMS-NSF Regional Conference Series in Applied Mathematics. https://doi.org/10.1137/1.9781611970265
82. Ravikiran, A.S.: What is Graph in Data Structure & Types of Graph? Jul 26, 2023. https://www.simplilearn.com/tutorials/data-structure-tutorial/graphs-in-data-structure
83. Graph Operations. https://pages.cs.wisc.edu/~paton/readings/Old/fall01/GRAPH-OPS.html
84. Basic Graph Theory. https://learn.saylor.org/mod/page/view.php?id=27304
85. Pandey, K.: Graph data structure - explained with examples, Jul 5, 2022. https://www.masaischool.com/blog/graph-data-structure-explained-with-examples/
86. Ford-Fulkerson Algorithm for Maximum Flow Problem. https://www.scaler.com/topics/data-structures/ford-fulkerson-algorithm-for-maximum-flow-problem/

Chapter 7
Network Centrality Solutions in SGL

7.1 Introduction

We are witnessing the rapidly growing use of networks and their different types, both physical and virtual, which have become an important part of our lives. Networking can be beneficial in both social and professional aspects. Social networking can be defined as using different social media platforms to connect with others who have similar interests. Networks are also created to satisfy objectives or needs. In today's competitive market, responsiveness to customer or supplier demand is often a decisive factor in the success of an organization, and the network is considered one of the most critical resources, both in private and public sectors.

Networks as models and operations are widely used in practically every area of human activity. In Chap. 2 we analyzed and reviewed very different types and areas of networking, which included: *Transport Networks* [1–6], *Communication Networks* [7–13], *Social Networks* [14–18], *Battle and Military Networks* [19–23], *Economic Networks* [24–29], *Distribution Networks* [30–34], *Virtual Networks* [35–39], *Neural Networks* [40–44], *Psychological Networks* [45–49], and *Criminal Networks* [50–55].

In Chap. 6, we reviewed publications on basic network and graph operations, including: *An introduction to networks* [56], *The degree distribution of a network* [57], *Random networks, Network algorithms* [58, 59], *Graph and network algorithms* [60], *Distributed network algorithms* [61], *Data structures and network algorithms* [62], *What is graph in data structure and types of graph* [63], *Graph operations* [64], *Basic graph theory* [65]. *Graph data structure* [66]. *Ford-Fulkerson algorithm for maximum flow problem* [67].

P. S. Sapaty, *Spatial Networking in the United Physical, Virtual, and Mental World*,
Studies in Systems, Decision and Control 533,
https://doi.org/10.1007/978-3-031-62154-3_7

The aim of this chapter is to investigate, evaluate, and model one of the most important features of graphs and networks, called ***Centrality***, with the use of networking-based Spatial Grasp Model and Technology [68–85]. In the graph analytics, Centrality is a very important concept in identifying nodes in a graph or network, and is used to measure the importance of various nodes, or how "central" they are in the graph. Centrality comes in different flavors and each flavor or a metric defines importance of a node from a different perspective and provides relevant analytical information about the graph and its nodes.

The rest of the chapter is organized as follows. Section 7.2 provides a brief summary on existing network centrality publications, reflecting: centrality, network centrality, centrality in social networks, centrality and underground network, centrality in psychological networks, centrality metrics for complex network analysis, and centrality in strategic transportation network design. Section 7.3 briefs the Spatial Grasp Model and Technology which allows for fully distributed and parallel operations on networked structures. Sections 7.4, 7.5, 7.6 and 7.7 describe how such network features as Degree Centrality, Closeness Centrality, Betweenness Centrality, and Eigen Vector Centrality can be expressed, implemented, and evaluated in SGL, with detailed examples of solutions provided. Section 7.8 concludes the paper. References reflect the provided reviews on networking publications and existing sources of the technology used.

7.2 The Concept of Network Centrality and Its Importance

We are offering a brief summary on the existing network centrality publications with their main ideas and applications.

Concepts of Centrality [86]. The importance of graph data structures depicting relationships among entities is growing. In graph analytics, Centrality is a very important concept in identifying nodes in a graph. It is used to measure the importance (or "centrality" as in how "central" a node is in the graph) of various nodes in a graph. Each node can be assessed from an angle depending on how "importance" is defined. Centrality comes in different flavors and each flavor or a metric defines importance of a node from a different perspective and further provides relevant information about the graph and its nodes.

Network Centrality [87]. Centrality is a key property of complex networks that influences the behavior of dynamical processes; it can bring important information about the organization of complex systems. There are many metrics to quantify the node centrality in networks. The paper reviews the main centrality measures and discusses their main features and limitations. The influence of network centrality on epidemic spreading and synchronization is also pointed out. Application of centrality measures to understand the function of complex systems is presented, including biological and cortical networks.

Centrality Measures in Social Networks [88]. Centrality measures are vital for understanding networks, also known as graphs. These algorithms use graph theory to calculate the importance of any given node in a network. They cut through noisy data, revealing parts of the network that need. The following centrality measures are popular. Degree centrality assigns an importance score based simply on the number of links held by each node. Betweenness centrality measures the number of times a node lies on the shortest path between other nodes. Closeness centrality scores each node based on their 'closeness' to all other nodes in the network.

Centrality and Underground Networks [89]. Novel centrality indices are defined. PathRank is a generalization of the PageRank algorithm, suitable to rank nodes of undirected graphs according to number and weight of paths in the graph. Icentr ranks nodes of the graph by means of a combination of the weights of nodes and edges, scaled according to the distance from each node, one at a time. The paper applies the two novel indices to underground transportation networks, since these networks represent an infrastructural backbone for the transportation system of most big cities over the world. A detailed study of Boston network is proposed.

Centrality in Psychological Networks [90]. Centrality indices are a popular tool to analyze structural aspects of psychological networks. As centrality indices were originally developed in the context of social networks, it is unclear to what extent these indices are suitable in a psychological network context. This article critically examines several issues with the use of the most popular centrality indices in psychological networks: degree, betweenness, and closeness centrality. It shows how problems with centrality indices discussed in the social network literature also apply to the psychological networks.

Centrality Metrics for Complex Network Analysis: Emerging Research and Opportunities [91]. It is a pivotal reference source for the latest research findings on centrality metrics and their broader applications for different categories of networks including wireless sensor networks, curriculum networks, social networks etc. Featuring extensive coverage on relevant areas, such as complex network graphs, node centrality metrics, and mobile sensor networks.

Centrality in Strategic Transportation Network Design [92]. Describes the appearance of a network by transportation network centrality, develops a strategic approach to transportation network design by conceptualizing transportation network centrality and relating it to the performance and quality of transportation networks. A practical implementation of this approach is provided, demonstrating its feasibility.

7.3 Basics of Spatial Grasp Model and Technology

Within Spatial Grasp Model and Technology (SGT), described in Chaps. 3, 4 and 5, also in [68–85], a high-level operational scenario expressed in recursive Spatial Grasp Language (SGL), starting in any world point *propagates, covers, and matches the distributed environment in parallel wavelike mode*, as symbolically shown in Fig. 7.1, Such propagation can result in returning and analyzing the reached states and data, which may be arbitrarily remote, or used for launching more waves.

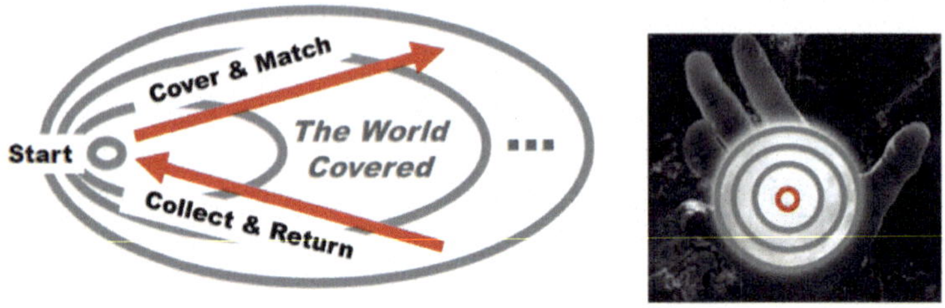

Fig. 7.1 Parallel wavelike world coverage under spatial grasp model

The SGL allows for organizing *direct space presence and operations* with unlimited powers and parallelism. Its universal recursive organization, with operational scenarios called *grasp*, can be expressed by a single formula:

$$\textbf{\textit{grasp}} \rightarrow \textbf{\textit{constant}} \mid \textbf{\textit{variable}} \mid \textbf{\textit{rule}}\ (\{\textbf{\textit{grasp}},\})$$

The *rule* expresses certain action, control, description or context accompanied with operands, which can be any *grasps* too. The rules, starting in certain points, can organize navigation of the world sequentially, in parallel, or any combinations thereof. They can result in the same application points or cause movement to other world points with obtained results left there or returned.

The *SGL interpreter* consists of specialized modules serving SGL scenarios or their parts happened to be inside this interpreter, also organizing exchanges with other interpreters for distributed SGL scenarios. Communicating SGL interpreters can be in arbitrary number of copies, representing *powerful spatial engines operating without central resources or control*, as in Fig. 7.2. They can effectively work with *spatial graph and network data* of any volumes and distributions. Details of SGT are explained in Chaps. 3, 4 and 5.

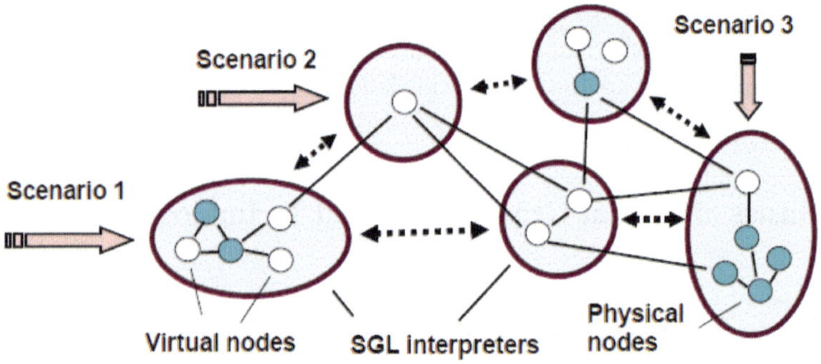

Fig. 7.2 SGL distributed networked interpretation

7.4 Degree Centrality in SGL

Degree of a node is defined as the number of direct connections a node has with other nodes. Mathematically, ***Degree Centrality*** for a node **i** can be defined as follows:

$$\mathrm{D(i)} = \sum_j m(i, j)$$

where m(i, j) = 1 if there is a link from node i to node j.

The following SGL scenario, applied in parallel to all network nodes of Fig. 7.3, will result with the node degrees shown at nodes in this figure.

```
output(hop(all_nodes);
    append(NAME, count(hop(all_neighbors))))
```

The printed output will be as follows:

(1:1, 2:7, 3:4, 4:5, 5:1, 6:2, 7:2, 8:3, 9:3, 10:2, 11:2, 12:2, 13:4, 14:2)

Or if shown as a table for convenience:

Nodes	1	2	3	4	5	6	7	8	9	10	11	12	13	14
Degree centrality	**1**	**7**	**4**	**5**	**1**	**2**	**2**	**3**	**3**	**2**	**2**	**2**	**4**	**2**

As can seen, node 2 is degree centrality strongest one (with seven direct neighbors), then go node 4 (with five), nodes 3 and 13 (with four), with the weakest being nodes 1 and 5 (with only one).

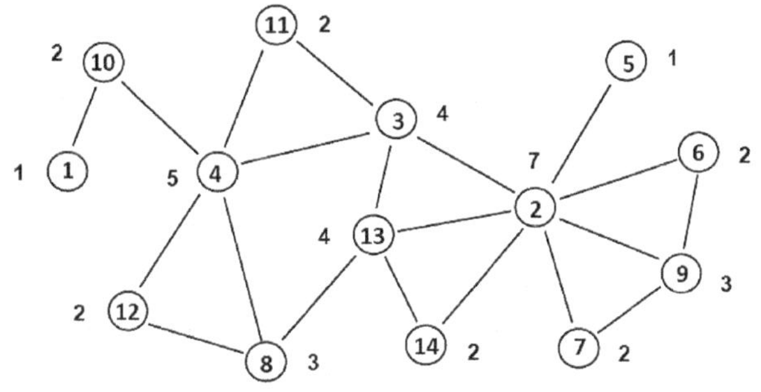

Fig. 7.3 Degree centrality of network nodes

7.5 Closeness Centrality in SGL

Geodesic distance from node *a* to node *b*, or *d(a, b)*, can be defined as the number of edges in the shortest path between *a* and *b* (i.e. with minimum number of edges), if a path exists between them. For the example shown in Fig. 7.4, the shortest paths between node 13 and all other nodes will be lying on the shortest path tree covering the whole network when starting from node 13.

To obtain the shortest path tree from node 13 embedded into the network body and output shortest distances from node 13 to all other nodes we may write the following SGL scenario, obtaining shortest distances from node 13 in other nodes in variables Distance, and finally printing all these distances together with the node names.

```
hop_node(13);
frontal(Far) = 0; nodal(Distance);
output(stay(repeat(hop(all_neighbors); Far += 1;
            or(Distance==nil, Distance>Far);
            Distance = Far));
     hop(all_other_nodes); append(NAME, Distance))
```

SGL output: (1:4, 2:1, 3:1, 4:2, 5:2, 6:2, 7:2, 8:1, 9:2, 10:3, 11:2, 12:2, 14:1).
In the form of a table this will be as follows:

Nodes	1	2	3	4	5	6	7	8	9	10	11	12	13	14
Distance from 13	4	1	1	2	2	2	2	1	2	3	2	2	0	1

Closeness Centrality C(i) of a node *i* in a graph can be defined by the sum of shortest paths from *i* to all other nodes *j*, as follows.

$$\mathrm{C(i)} = \sum_{j} d(i,j)$$

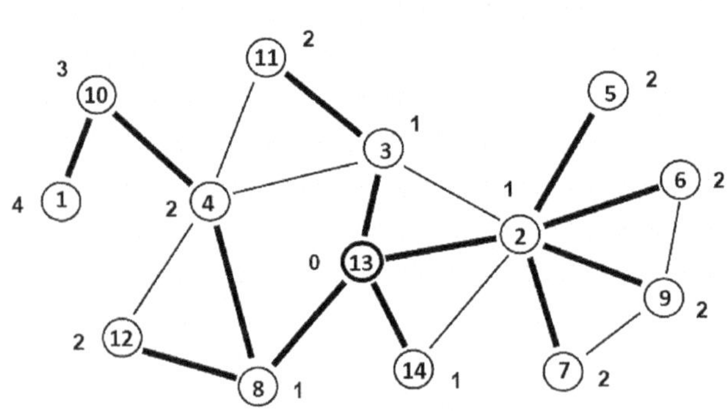

Fig. 7.4 Example of shortest paths from node 13 to other nodes with distances to them

Distances to all other nodes for each node and their summary can be found by the following SGL scenario (see full results of its operation in Table 7.1).

```
output(
 hop(all_nodes); IDENTITY = NAME;
   frontal(Far) = 0; nodal(Distance);
   stay(repeat(hop(all_neighbors); Far += 1;
         or(Distance==nil, Distance>Far);
         Distance = Far));
   append(NAME,
       sum(hop(all_other_nodes); Distance)))
```

The above scenario starts in parallel from all nodes of the network and then creates from each node an individual shortest path tree covering the whole network (using for each tree a unique identity corresponding to the name of the node it started, not to interfere between the trees created in parallel). After obtaining the trees, embedded into the network body, the scenario then hops from each node to all other nodes and summarizes distances to them registered in its individual tree (as in Table 7.1). After this it collects in parallel and outputs all node names with summaries of distances to other nodes (as in the rightmost column of Table 7.1).

The printed node closeness centrality result reflecting the sum of distances to all other nodes will be as follows:

(1:48, 2:23, 3:23, 4:26, 5:35, 6:34, 7:34, 8:28,
9:33, 10:36, 11:30, 12:35, 13:25, 14:32)

As a table this may be expressed as:

Table 7.1 Closeness from each node to all other nodes and their summary

↱	1	2	3	4	5	6	7	8	9	10	11	12	13	14	SUM
1	0	4	3	2	5	5	5	3	5	1	3	3	4	5	**48**
2	4	0	1	2	1	1	1	2	1	3	2	3	1	1	**23**
3	3	1	0	1	2	2	2	2	2	2	1	2	1	2	**23**
4	2	2	1	0	3	3	3	1	3	1	1	1	2	3	**26**
5	5	1	2	3	0	2	2	3	2	4	3	4	2	2	**35**
6	5	1	2	3	2	0	2	3	1	4	3	4	2	2	**34**
7	5	1	2	3	2	2	0	3	1	4	3	4	2	2	**34**
8	3	2	2	1	3	3	3	0	3	2	2	1	1	2	**28**
9	5	1	2	3	2	1	1	3	0	4	3	4	2	2	**33**
10	1	3	2	1	4	4	4	2	4	0	2	2	3	4	**36**
11	3	2	1	1	3	3	3	2	3	2	0	2	2	3	**30**
12	3	3	2	1	4	4	4	1	4	2	2	0	2	3	**35**
13	4	1	1	2	2	2	2	1	2	3	2	2	0	1	**25**
14	5	1	2	3	2	2	2	2	2	4	3	3	1	0	**32**

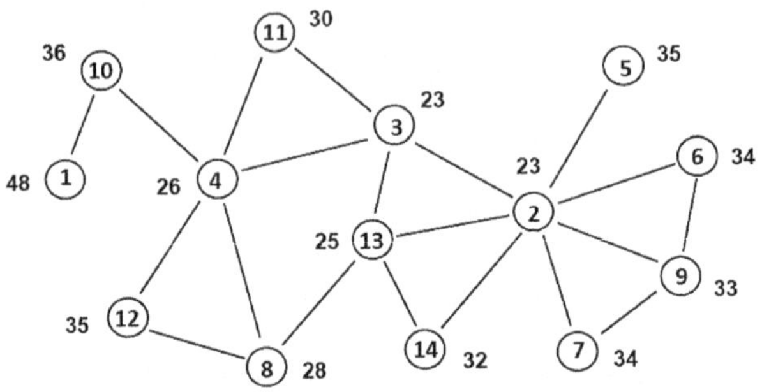

Fig. 7.5 Closeness centrality of network nodes

Nodes	1	2	3	4	5	6	7	8	9	10	11	12	13	14
Closeness centrality	**48**	**23**	**23**	**26**	**35**	**34**	**34**	**28**	**33**	**36**	**30**	**35**	**25**	**32**

See also Fig. 7.5, with closeness centrality values associated with nodes.

As can be seen, the strongest closeness centrality nodes are 2 and 3 (with same sum of distances 23), then go node 13 (with 25), node 4 (with 26), and the weakest being node 1 (with 48).

7.6 Betweenness Centrality in SGL

This metric defines and measures the importance of a node in a network based upon how many times it occurs in the shortest path between all pairs of nodes in a graph. Mathematically, ***Betweenness Centrality B(i)*** of a node *i* in a graph is defined as follows:

$$\mathrm{B(i)} = \sum_{a,b} \frac{g_{aib}}{g_{ab}}.$$

where *a*, *b* is any pair of nodes in the graph, g_{aib} is the number of shortest paths from node *a* to node *b* passing through *i*, and g_{ab} is the number of shortest paths from *a* to *b*. An example is shown in Fig. 7.6 where nodes 10, 4, 3, and 2 are on the shortest path between nodes 1 and 9, which can be found by the SGL scenario below.

```
hop_node(1);
frontal(Far) = 0; nodal(Distance, Prev);
stay(repeat(hop(all_neighbors); Far += 1;
        or(Distance==nil, Distance>Far);
        Distance = Far; Prev = PREVIOUS));
hop_node(9); repeat(hop(Prev); nonequal(NAME, 1);
output(NAME))
```

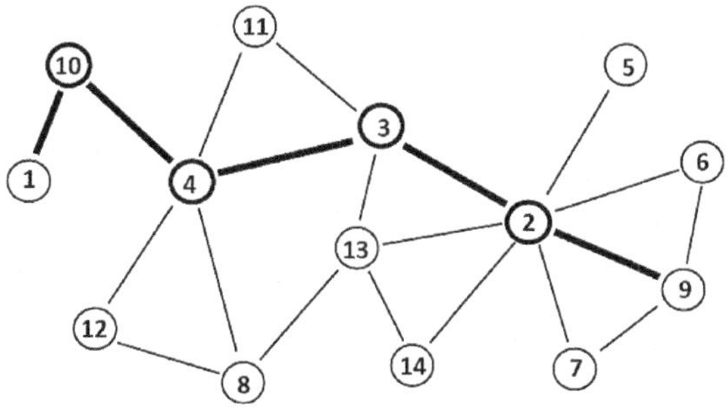

Fig. 7.6 Example of nodes staying on the shortest path between nodes 1 and 9

SGL output will be: (2, 3, 4, 10).

The following SGL scenario will be finding and summarizing betweenness for all nodes lying on shortest paths between pairs of all other nodes

```
sequence(
  (hop(all_nodes); IDENTITY = NAME;
    frontal(Far) = 0; nodal(Distance, Prev);
    stay(repeat(hop(all_neighbors); Far += 1;
           or(Distance==nil, Distance>Far);
           Distance = Far; Prev = PREVIOUS));
    hop(all_other_nodes);
    repeat(hop(Prev); nonequal(NAME, IDENTITY);
        CONTENT += 1))),
output(hop(all_nodes); append(NAME, CONTENT)))
```

This scenario starts in parallel from all nodes of the network and then creates from each node an individual shortest path tree covering the whole network (using for each tree a unique identity corresponding to the name of the node it started, not to interfere between the trees created in parallel). After obtaining the trees, embedded into the network body, the scenario hops from each node to all other nodes and repetitively propagates up the recorded individual shortest path tree with incrementing the neutral counter (as content) of each passed node (except the starting ones).

After completion of this multiple trees parallel ascending process for all nodes, the scenario then outputs the node names with their resultant betweenness summaries, accumulated as contents. Examples of its operation when repeatedly starting from the node on the left and leading to all the nodes above are shown in Table 7.2 (from nodes 1 and 14 only, with local summaries below, similarly from others), and the resultant global betweenness summaries on all nodes are provided in Table 7.3.

The printed SGL result on the betweenness values for all nodes will be as follows:

(1:0, 2:72, 3:70, 4:60, 5:0, 6:0, 7:0, 8:22, 9:2, 10:24, 11:0, 12:0, 13:33, 14:0)

In a more convenient form this may be expressed as follows:

Table 7.2 Nodes staying on shortest paths from a certain node to all other nodes and their summary: (a) From node 1, (b) From node 14

↱	1	2	3	4	5	6	7	8	9	10	11	12	13	14
1		○	1	1						1				
1			○	1						1				
1				○						1				
1		1	1	1	○					1				
1		1	1	1		○				1				
1		1	1	1			○			1				
1				1				○		1				
1		1	1	1					○	1				
1										○				
1				1						1	○			
1				1						1		○		
1			1	1						1			○	
1			1	1						1			1	○
F1		4	7	11						12			1	

a

...

↱	1	2	3	4	5	6	7	8	9	10	11	12	13	14
14	○		1	1				1		1			1	
14		○												
14		1	○										1	
14			1	○				1					1	
14		1			○									
14		1				○								
14		1					○							
14								○					1	
14		1							○					
14			1	1				1		○			1	
14			1								○		1	
14								1				○	1	
14													○	
F14		5	4	2				4		1			7	

b

Table 7.3 Resultant betweenness summary from the local summaries on all nodes

	1	2	3	4	5	6	7	8	9	10	11	12	13	14
F1		4	7	11						12			1	
F2			5	3				1		1			2	
F3		5		4						1			2	
F4		4	7					2		1			1	
F5		12	5	3				1		1			2	
F6		12	5	3				1	1	1			2	
F7		6	4	3				1	1	1			1	
F8		4		3						1			6	
F9		9	7	3				1		1			1	
F10		4	7	11				2					1	
F11		4	7	4						1			1	
F12		3	4	8				7		1			6	
F13			8	2				2		1				
F14		5	4	2				4		1			7	
FIN		72	70	60				22	2	24			33	

Nodes	1	2	3	4	5	6	7	8	9	10	11	12	13	14
Betweenness centrality	**0**	**72**	**70**	**60**	**0**	**0**	**0**	**22**	**2**	**24**	**0**	**0**	**33**	**0**

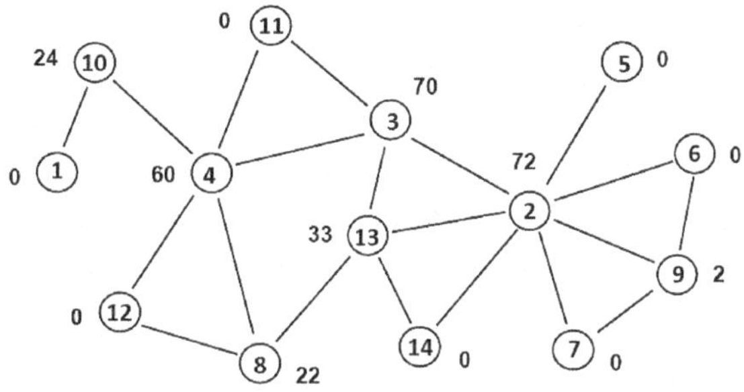

Fig. 7.7 Betweenness centrality of network nodes

See also Fig. 7.7 where betweenness values are shown associated with the related network nodes.

As can be seen, the strongest betweenness centrality nodes are 2 (72) and 3 (70), then goes node 4 (60), whereas the weakest being nodes 1, 5, 6, 7, 11, 12, and 14 (just 0).

7.7 Eigen Vector Centrality in SGL

The last centrality feature considered in this paper is known as the **Eigen Vector Centrality**. This metric measures the importance of a node in a graph as a function of the importance of its neighbors. If a node is connected to highly important nodes, it will have a higher Eigen Vector Centrality score as compared to a node which is connected to lesser important nodes. The following SGL scenario, starting from the degree centrality of each node (i.e. the number of its direct neighbors, as in Fig. 7.3), calculates importance of each node as the sum of centrality values of all its direct neighbors.

```
nodal(Old, New);
sequence(
  (hop(all_nodes); Old = count(hop(all_neighbors)),
  output(hop(all_nodes;
       New = count(hop(all_neighbors); Old);
      append(NAME, New)))
```

This scenario first accesses all network nodes in parallel with assigning them their degree centrality value (as an "old" one, by the number of their direct neighbors), and then reassigns to each of them the new value as a sum of previous, or old, values of their direct neighbors, considering new values as their Eigen ones, outputting the latter. The printed SGL Eigen result will be as follows:

(1:2, 2:18, 3:18, 4:13, 5:7, 6:10, 7:10, 8:11, 9:11, 10:6, 11:9, 12:8, 13:16, 14:11)

In a more convenient form (Eigen values shown in red) this will be as follows (see also Fig. 7.8).

Nodes:	1	2	3	4	5	6	7	8	9	10	11	12	13	14
Eigen Vector Centrality:	2	18	18	13	7	10	10	11	11	6	9	8	16	11

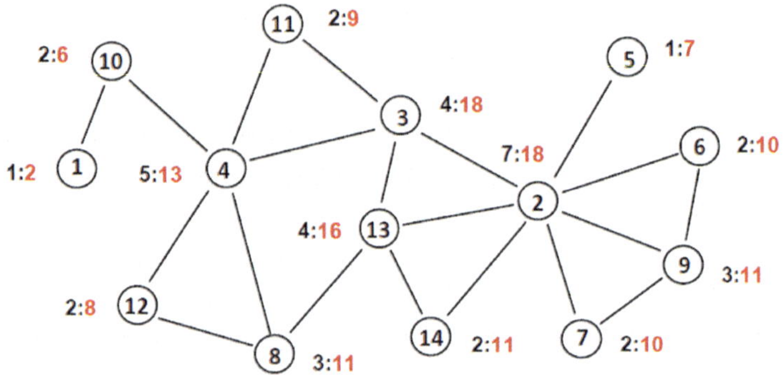

Fig. 7.8 Eigen centrality of all nodes, step 1

As can be seen, the strongest Eigen Centrality nodes are 2 and 3 (both with value 18), then go nodes 13 (with 16), and 4 (with 13).

This finding of new node values as a summary of previous values of all neighboring nodes may be repeated, each time receiving more global Eigen values of nodes, as follows.

```
nodal(Old, New);
sequence(
  (hop(all_nodes); Old = count(hop(all_neighbors))),
  loop(2)(output(align(hop(all_nodes;
        New = count(hop(all_neighbors); Old));
        Old = New; append(NAME, New)))))
```

For the two repetitions of the above scenario we will receive the following summary (see the table below, also Fig. 7.9, with the second step Eigen values shown in blue).

As can be seen, after the second iteration the strongest Eigen Centrality node is 2 (with value 83), then go node 13 (with 58), node 3 (with 56), and node 4 (with 52).

7.8 Conclusions

The results of this chapter confirm availability and efficiency of using SGT and SGL for investigating, expressing, and evaluating such important network feature as nodes Centrality, with such variants as Degree Centrality, Closeness Centrality, Betweenness Centrality, and Eigen Vector Centrality. SGL allows us to provide detailed, very clear and extremely compact solutions for the centrality problems which can operate on arbitrary large and complex networks, and in highly parallel and fully distributed mode. The solutions provided and discussed in this chapter were obtained in a spatial thinking, spatial pattern recognition, and spatial pattern matching mode, rather than in traditional logic-based and algorithmic philosophy and culture. Moreover, being

Nodes:	1	2	3	4	5	6	7	8	9	10	11	12	13	14
Eigen Vector Centrality:	2	18	18	13	7	10	10	11	11	6	9	8	16	11
Eigen Vector Centrality:	6	83	56	52	18	29	29	37	38	15	31	24	58	34

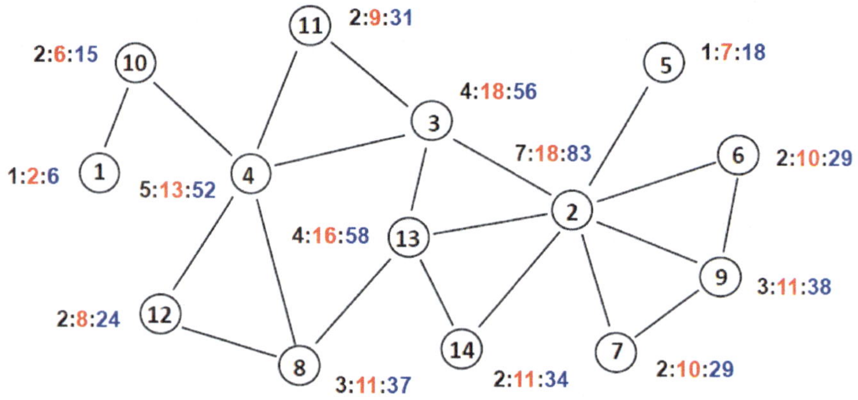

Fig. 7.9 Eigen centrality of all nodes, two steps

highly parallel and fully distributed, they challenge existing opinions that parallel and distributed algorithms are usually more complex than traditional sequential ones for solving the same problems, just proving the opposite, and especially for the network-related operations.

References

1. Transport Networks: Inspire Registry (2015). https://inspire.ec.europa.eu/theme/tn
2. Transportation Networks: Travel Forecasting Resource. https://tfresource.org/topics/Transportation_networks.html
3. Durkin, J., Goodman, J. et al.: Building Multiservice Transport Networks (Networking Technology), 1st edn. Cisco Press (2012)
4. Murakami, M., Li, H., Ryoo, J.-d.: Packet Transport networks: Overview and future direction. In: 2nd APT/ITU Conformance and Interoperability Workshop (C&I-2) (2014, August). https://www.itu.int/en/ITU-T/C-I/interop/Documents/20140826/CI-2_INP-09_Packet_Transport_Networks_Overview_and_Future_Direction.pdf
5. Air Transport Network: https://en.wikipedia.org/wiki/Air_Transport_Network
6. Qureshi, K.N., Ul Islam, F. et al.: Improved road segment-based geographical routing protocol for vehicular ad-hoc networks. Electronics **9**(8), 1248 (2020). https://doi.org/10.3390/electronics9081248. https://www.mdpi.com/2079-9292/9/8/1248
7. Sujan: What is Communication Network? Definition, Importance, and 4 Types (Explained) (2023, June 24). https://tyonote.com/communication_network/

8. Fitzek, F.H.P., Granelli, F. et al.: Computing in Communication Networks: From Theory to Practice, 1st edn. Academic Press (2020, June 4)
9. Srikant, R., Ying, L.: Communication Networks: An Optimization, Control, and Stochastic Networks Perspective, 1st edn. Cambridge University Press (2013, November 14)
10. Raghunandan, K.: Introduction to Wireless Communications and Networks: A Practical Perspective (Textbooks in Telecommunication Engineering), 1st ed. 2022 edn. Springer (2023, April 2)
11. Lou, Y., Ahmed, N.: Underwater Communications and Networks (Textbooks in Telecommunication Engineering). Springer (2021, November 15)
12. Kaufman, C., Perlman, R. et al.: Network Security: Private Communication in a Public World (Prentice Hall Series in Computer Networking and Distributed Systems), 3rd edn. Addison-Wesley Professional (2022, September 26)
13. Wu, Y., Singh, S. et al.: 6G Mobile Wireless Networks (Computer Communications and Networks), 1st ed. 2021 edn. Springer (2022, August 26)
14. Social Network: https://en.wikipedia.org/wiki/Social_network
15. Kenton, W.: What Is Social Networking? Investopedia. https://www.investopedia.com/terms/s/social-networking.asp#:~:text=Social%20networking%20refers%20to%20using,)%2C%20Instagram%2C%20and%20Pinterest. Accessed 24 May 2023
16. Nanawati, S.: Social Network Analytics, Published in Analytics Vidhya (2019, August 1). https://medium.com/analytics-vidhya/social-network-analytics-f082f4e21b16
17. Singh, C.: 20+ Most Popular Social Networking Sites in 2024. (2023, December 17). https://www.socialpilot.co/social-networking-sites
18. Jackson, M.O.: Social and Economic Networks, Illustrated edn. Princeton University Press (2010, November)
19. Harrison, T.: Battle Networks and the Future Force, Center for Strategic and International Studies (2021). https://www.csis.org/analysis/battle-networks-and-future-force
20. Stillion, J., Clark, B.: What It Takes To Win: Succeeding in 21st Century Battle Network Conmpetitions. Center for Strategic and Budgetary Assessments (CSBA) (2015). https://csbaonline.org/uploads/documents/What-it-Takes-to-Win.pdf
21. Hadley, G.: DAF outlines a new 'Battle Network' as its contribution to JADC2. Air & Space Forces Magazine (2023, March 23). https://www.airandspaceforces.com/daf-battle-network-contribution-jadc2/
22. The challenges of military networks and technological developments. Elbit Systems. https://elbitsystems.com/e-lynx_multi_channel/
23. Themnér, A., Karlén, N.: Building a safety net: explaining the strength of ex-military networks. Secur. Stud. **29**(2), 268–300 (2020). https://doi.org/10.1080/09636412.2020.1722851
24. What is an economic network? Advantages and disadvantages of economic networks. Vietnambiz (2019). https://vietnambiz.vn/mang-luoi-kinh-te-economic-network-la-gi-uu-nhuoc-diem-cua-mang-luoi-kinh-te-20191011113428212.htm
25. Halton, C.: economic network: what it is, pros and cons. Examples. https://www.investopedia.com/terms/e/economic-network.asp#:~:text=Key%20Takeaways-,An%20economic%20network%20is%20a%20combination%20of%20individuals%2C%20groups%2C%20or,companies%20or%20partnerships%20between%20corporations. Accessed 24 Oct 2022
26. Todo, Y., Matous, P.: Economic networks for more innovative and resilient economies (2015, 26 Junuary). https://cepr.org/voxeu/columns/economic-networks-more-innovative-and-resilient-economies
27. Emmert-Streib, F., Tripathi, S. et al.: Understanding the world economy in terms of networks: a survey of data-based network science approaches on economic networks. Front. Appl. Math. Stat. (Sec. Mathematical Finance) **4** (2018). https://doi.org/10.3389/fams.2018.00037
28. Wixted, B.: Innovation System Frontiers: Cluster Networks and Global Value (Advances in Spatial Science). Springer (2009)
29. Manyika, J.: What the economy looks like so far for individuals in the 21st century (2020, February 6). https://www.linkedin.com/pulse/what-economy-looks-like-so-far-individuals-21st-century-james-manyika

30. Hayes, A.: Distribution network: definition, how it works, and examples. Investopedia. https://www.investopedia.com/terms/d/distribution-network.asp#:~:text=In%20a%20supply%20chain%2C%20a,or%20through%20a%20retail%20network. Accessed 22 February 2021
31. Distribution Network: The flow of goods from a producer or supplier to an end consumer. Written by CFI Team. https://corporatefinanceinstitute.com/resources/valuation/distribution-network/
32. What are types of distribution networks in supply chain management? oboloo FAQ's. https://oboloo.com/blog/what-are-types-of-distribution-networks-in-supply-chain-management/
33. Distribution network—what is it and what are the benefits? https://www.salesbabu.com/blog/what-is-a-distribution-network-and-its-benefits/
34. Wu, Q., Shen, F. et al.: Optimal Operation of Active Distribution Networks: Congestion Management, Voltage Control and Service Restoration. Academic Press (2023)
35. Watts, S.: What is a virtual network? BMC (2020, June 16). https://www.bmc.com/blogs/virtual-network/
36. What is virtual networking? VMware by Broadcom. https://www.vmware.com/topics/glossary/content/virtual-networking.html
37. Network virtualization. https://en.wikipedia.org/wiki/Network_virtualization.
38. Carlos, O., Duarte, M.B., Pujolle, G.: Virtual Networks: Pluralistic Approach for the Next Generation of Internet (Networks and Telecommunications), 1st edn. Wiley-ISTE (2013, August 2).
39. Fundamentals of virtual networking. https://www.geeksforgeeks.org/fundamentals-of-virtual-networking/
40. Neural network. https://en.wikipedia.org/wiki/Neural_network
41. Anand, V.: Introduction to Neural Network in Machine Learning. https://www.analyticsvidhya.com/blog/2022/01/introduction-to-neural-networks/. Accessed 10 August 2023
42. Ashtari, H.: What is a neural network? definition, working, types, and applications in 2022. https://www.spiceworks.com/tech/artificial-intelligence/articles/what-is-a-neural-network/. Accessed 3 August 2022
43. Artificial Neural Network: NVIDIA Developer. https://developer.nvidia.com/discover/artificial-neural-network
44. Wu, L., Cui, P. et al.: Graph Neural Networks: Foundations, Frontiers, and Applications, 1st ed. 2022 edn. Springer (2022, January 4)
45. Borsboom, D., Deserno, M.K., Rhemtulla, M. et al.: Network analysis of multivariate data in psychological science. Nat. Rev. Methods Prim. **1**, 58 (2021). https://doi.org/10.1038/s43586-021-00055-w. https://www.nature.com/articles/s43586-021-00055-w
46. Jones, P.J., Mair, P., McNally, R.J.: Visualizing psychological networks: a tutorial in R. Front. Psychol. (Sec. Quantitative Psychology and Measurement) **9** (2018). https://doi.org/10.3389/fpsyg.2018.01742
47. Fonseca-Pedrero, E.: Network analysis in psychology. Psychol. Pap. **39**(1), 1–12 (2018). https://doi.org/10.23923/pap.psicol2018.2852. https://www.papelesdelpsicologo.es/pdf/2852.pdf
48. Bringmann, L.F.: Person-specific networks in psychopathology: past, present, and future. Curr. Opin. Psychol. **41**, 59–64 (2021). https://www.sciencedirect.com/science/article/pii/S2352250X21000300?via%3Dihub
49. Yuqing, C., Shuyang, D. et al.: Network analysis and its applications in psychology. Adv. Psychol. Sci. **28**(1), 178–190 (2020). https://doi.org/10.3724/SP.J.1042.2020.00178
50. Campana, P.: Explaining criminal networks: Strategies and potential pitfalls. Methodol. Innov. **9**, 1–10 (2016). https://doi.org/10.1177/2059799115622748
51. Bichler, G.: Understanding Criminal Networks: A Research Guide, 1st edn. University of California Press (2019, September 24)
52. Morselli, C.: Crime and Networks (Criminology and Justice Studies). Routledge (2013)
53. New forms of organized crime: networked structure. UNODS, Unite Nations Office on Drugs and Crime. https://www.unodc.org/e4j/en/organized-crime/module-7/key-issues/networked-structure.html

54. Hufnagel, S., Moiseienko, A.: Criminal Networks and Law Enforcement: Global Perspectives on Illegal Enterprise (Transnational Criminal Justice), 1st edn. Routledge (2019, June 27)
55. Detecting and disrupting transnational criminal organizations: analytics for interdependent smuggling and money-laundering networks. Criminal Investigation and Network Analysis, DHS Centers of Excellence. https://cina.gmu.edu/projects/detecting-and-disrupting-transnational-criminal-organizations-analytics-for-interdependent-smuggling-and-money-laundering-networks/
56. An Introduction to Networks. https://mathinsight.org/network_introduction
57. The Degree Distribution of a Network. https://mathinsight.org/degree_distribution
58. Random Networks. https://mathinsight.org/random_network
59. Network Algorithms. https://ics.uci.edu/~goodrich/teach/cs165/notes/NetworkAlgs.pdf
60. Graph and Network Algorithms. https://ch.mathworks.com/help/matlab/graph-and-network-algorithms.html
61. Schmid, S., Mandal, P.S.: Distributed Network Algorithms (Lecture Notes for GIAN Course) (2016, Summer). https://schmiste.github.io/GIAN-Lecture-Notes-NetAlg.pdf
62. Tarjan, R.E.: Data structures and network algorithms. In: CBMS-NSF Regional Conference Series in Applied Mathematics. https://doi.org/10.1137/1.9781611970265
63. Ravikiran, A.S.: What is graph in data structure & types of graph? (2023, July 26). https://www.simplilearn.com/tutorials/data-structure-tutorial/graphs-in-data-structure
64. Graph Operations. https://pages.cs.wisc.edu/~paton/readings/Old/fall01/GRAPH-OPS.html
65. Basic Graph Theory. https://learn.saylor.org/mod/page/view.php?id=27304
66. Pandey, K.: Graph Data Structure—Explained With Examples (2022, July 5). https://www.masaischool.com/blog/graph-data-structure-explained-with-examples/
67. Ford-Fulkerson Algorithm for Maximum Flow Problem. https://www.scaler.com/topics/data-structures/ford-fulkerson-algorithm-for-maximum-flow-problem/
68. Sapaty, P.S.: A distributed processing system. European Patent N 0389655, Publ. 10.11.93, European Patent Office
69. Sapaty, P.S.: Mobile Processing in Distributed and Open Environments. John Wiley & Sons, New York (1999)
70. Sapaty, P.S.: Ruling Distributed Dynamic Worlds. John Wiley & Sons, New York (2005)
71. Sapaty, P.S.: Managing Distributed Dynamic Systems with Spatial Grasp Technology. Springer (2017)
72. Sapaty, P.S.: Holistic Analysis and Management of Distributed Social Systems. Springer (2018)
73. Sapaty, P.S.: Complexity in International Security: A Holistic Spatial Approach. Emerald Publishing (2019)
74. Sapaty, P.S.: Symbiosis of Real and Simulated Worlds under Spatial Grasp Technology. Springer (2021)
75. Sapaty, P.S.: Spatial Grasp as a Model for Space-based Control and Management Systems. CRC Press (2022)
76. Sapaty, P.S.: The Spatial Grasp Model: Applications and Investigations of Distributed Dynamic Worlds. Emerald Publishing (2023)
77. Sapaty, P.S.: Providing Integrity, Awareness, and Consciousness in Distributed Dynamic Systems. CRC Press (2025)
78. Sapaty, P.S.: Spatial management of air and missile defence operations. Math. Mach. Syst. 1 (2023). http://www.immsp.kiev.ua/publications/articles/2023/2023_1/01_23_Sapaty.pdf
79. Sapaty, P.S.: Providing distributed system integrity under spatial grasp technology. Math. Mach. Syst. 2 (2023). http://www.immsp.kiev.ua/publications/articles/2023/2023_2/02_23_Sapaty.pdf
80. Sapaty, P.S.: Providing global awareness in distributed dynamic systems. Int. Relat. Dipl **11**(2), 87–100 (2023). https://doi.org/10.17265/2328-2134/2023.02.002. https://www.davidpublisher.com/Public/uploads/Contribute/6486c3d05a6cc.pdf
81. Sapaty, P.S.: Simulating distributed consciousness with spatial grasp model. Math. Mach. Syst. 3 (2023)

82. Sapaty, P.S.: Network centrality operations under spatial grasp technology. J. Adv. Artif. Intell. Mach. Learn. **1**(1), 1–11 (2023). https://www.scivisionpub.com/journals/articleinpress-journal-of-advances-in-artificial-intelligence-and-machine-learning
83. Sapaty, P.S.: Managing distributed systems with spatial grasp patterns. Math. Mach. Syst. 4 (2023)
84. Sapaty, P.S.: Spatial networks as models for organoid cultures and brain research. Math. Mach. Syst. 1 (2024)
85. Sapaty, P.S.: Basic network operations under spatial grasp model. Sci. Set J. Phys. **3**(1), 01–16 (2024)
86. Bhasin, J.: Graph analytics—introduction and concepts of centrality. Towards Data Science (2019, August 14). https://towardsdatascience.com/graph-analytics-introduction-and-concepts-of-centrality-8f5543b55de3
87. Rodrigues, F.A.: Network centrality: an introduction (2019, January 22). arXiv preprint arXiv: 1901.07901v1. https://arxiv.org/pdf/1901.07901.pdf
88. Disney, A.: Social network analysis 101: centrality measures explained. Cambridge Intelligence (2020, January). https://cambridge-intelligence.com/keylines-faqs-social-network-analysis/
89. Mussone, L., Viseh, H., Notari, R.: Novel centrality measures and applications to underground networks. Phys. A: Stat. Mech. Appl. **589**, 126595 (2022). https://www.sciencedirect.com/science/article/abs/pii/S0378437121008633
90. Bringmann, L.F., Elmer, T., Epskamp, S., Krause, R.W. et al.: What do centrality measures measure in psychological networks? J. Abnorm. Psychol. **128**(8), 892–903 (2019). https://doi.org/10.1037/abn0000446. 0021-843X/19/$12.00. https://www.apa.org/pubs/journals/features/abn-abn0000446.pdf
91. Meghanathan, N.: Centrality Metrics for Complex Network Analysis: Emerging Research and Opportunities (Advances in Wireless Technologies and Telecommunication). IGI Global (2018). https://www.igi-global.com/book/centrality-metrics-complex-network-analysis/181520#table-of-contents
92. Lange, A.: Centrality in Strategic Transportation Network Design: An application to Less-Than-Truckload Networks (Edition KWV). Springer Gabler (2019)

Chapter 8
Examples of Networking Solutions in Combined Worlds

8.1 Introduction

The chapter aims at development of effective methods capable of integrating very different types of networks into higher-level infrastructures representing advanced social systems, particularly considering symbiosis of goods production, goods evaluation, and obtained products delivery systems under some universal networking concept.

In Chap. 2 we analyzed and reviewed very different types and areas of networking, which included: *Transport Networks* [1–6], *Communication Networks* [7–13], *Social Networks* [14–18], *Battle and Military Networks* [19–23], *Economic Networks* [24–29], *Distribution Networks* [30–34], *Virtual Networks* [35–39], *Neural Networks* [40–44], *Psychological Networks* [45–49], and *Criminal Networks* [50–55].

In Chap. 6, we analyzed how basic operations on networks can be performed by SGT, having reviewed the following sources: *An introduction to networks* [56], *The degree distribution of a network* [57], *Random networks, Network algorithms* [58, 59], *Graph and network algorithms* [60], *Distributed network algorithms* [61], *Data structures and network algorithms* [62], *What is graph in data structure and types of graph* [63], *Graph operations* [64], *Basic graph theory* [65], *Graph data structure* [66], and *Ford-Fulkerson algorithm for maximum flow problem* [67].

In Chap. 7, network centrality operations were investigated by SGT too, after reviewing the following sources: *Concepts of Centrality* [68], *Network Centrality* [69], *Centrality Measures in Social Networks* [70], *Centrality and Underground Networks* [71], *Centrality in Psychological Networks* [72], *Centrality Metrics for Complex Network Analysis* [73], and *Centrality in Strategic Transportation Network Design* [74].

After this analysis we have come to the conclusion of suitability and efficiency of both theoretical and practical use of the same Spatial Grasp Model and Technology (SGT) with its basic Spatial Grasp Language (SGL) [75–92], also detailed in Chaps. 3,

P. S. Sapaty, *Spatial Networking in the United Physical, Virtual, and Mental World*, Studies in Systems, Decision and Control 533,
https://doi.org/10.1007/978-3-031-62154-3_8

4 and 5, for expressing and solving numerous problems in very different types of networks.

The *goal of this chapter* is to investigate applicability of SGT and SGL for the development of higher-level social infrastructures effectively ***integrating very different types of networking***, like reflecting distributed goods production, different items evaluation and assessment by consumers, and requested goods delivery to the registered users. The rest of it is organized as follows.

Section 8.2 describes and reviews publications on networking in the following areas: *production networks*, *product evaluation networks*, and *product distribution networks*. Section 8.3 briefs main features of SGT, SGL, and its distributed networked implementation, useful for better comprehension of the following sections. Section 8.4 provides practical examples of combined networking solutions in SGL for: (1) Product development networks, expressing design and production of different type products also possible links between these networks, and finding their strongest or central nodes; (2) Product assessment and request networks, finding strongest communities interested in same or different products like discovering cliques among them; (3) Networks of product delivery to most interested consumers, with delivery of resultant products via specially designed optimal infrastructures; and (4) Summary of these combined networking solutions. Section 8.5 concludes the paper confirming effectiveness of SGT for various network operations, including higher-level ones integrating different types of networks within advanced social infrastructures. References include many publication sources which were reviewed, analyzed, classified, and used in this paper.

8.2 Examples of Networking in the Selected Areas

The following is a brief analysis and review of different types of networks chosen for investigation of their possible holistic integration within higher-level social infrastructures, which may be both national and international.

- **Production Networks**

A Global Production Network (GPN) [93] is one whose interconnected nodes and links extend spatially across national boundaries and, in so doing, integrates parts of disparate national and sub-national territories. GPN frameworks combines the insights from the global value chain analysis and actor–network theory.

Design and operation of industrial companies acting in global production networks is discussed in [93]. Globalization has led to profound changes in the economy in recent decades. Nowadays, companies of any size operate globally in the form of GPNs.

Centers and connectivity in global production networks are discussed in [94]. Global production networks fluxes go beyond materials, energy or other tangible goods. It has impacts on the movement of ideas, information, culture and people.

See Fig. 8.1 symbolically related to production networks.

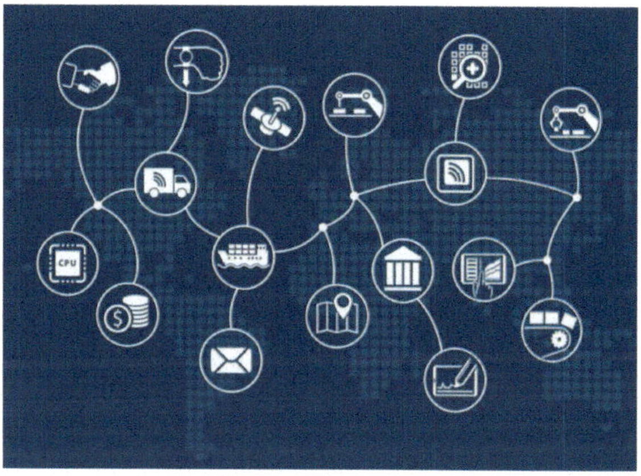

Fig. 8.1 Production networks

- **Product Evaluation Networks**

Evaluation of product development systems by network analysis is in [95]. It proposes the integration of two systems engineering analysis tools, the Design Structure Matrix (DSM) and Network Analysis (NA), to study task interactions in a Product Development Process (PDP).

Sustainable product design evaluation based on deep residual networks is presented in [96]. Sustainable product design evaluation has become an increasingly crucial aspect of product design. This study proposes an integrated approach that combines manual design evaluation based on the analytic hierarchy process (AHP) with an automatic design evaluation.

Machine learning and network algorithms for expressing product opinions are in [97]. E-commerce is the fastest-growing segment of the economy. Online reviews play a crucial role in helping consumers evaluate and compare products and services. There are many reasons why it is hard to identify opinion spammers automatically, including the absence of reliable labeled data.

See Fig. 8.2 symbolically related to product assessment networks.

- **Product Distribution Networks**

The role of decision-making parameters for distribution networks is discussed in [98]. Distribution networks have to interconnect the consumption and production sides of the supply chain. Decision making has special parameters according to perspectives related to strategic, tactical and operational levels.

A value delivery network in marketing [99] refers to the network of relationships and interactions between various entities involved in delivering value to customers. It includes the coordination and collaboration between suppliers, manufacturers, distributors, and customers to ensure the effective and efficient delivery of products or services.

A distribution network can be seen as the flow of goods from a producer or supplier to an end consumer [100]. The network consists of storage facilities, warehouses,

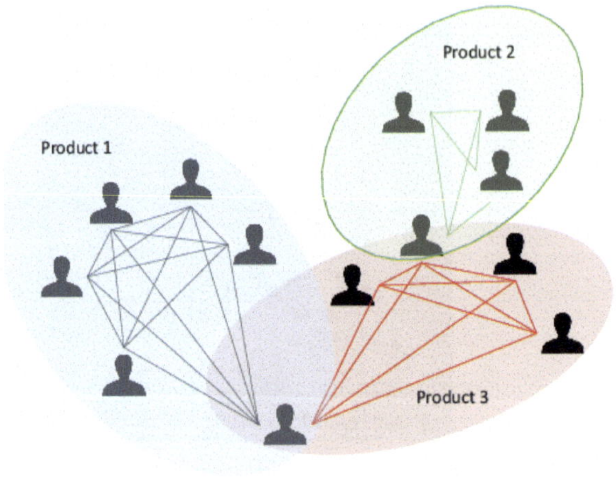

Fig. 8.2 User interconnections for assessment of different products

Fig. 8.3 Products distribution networks

and transportation systems that support the movement of goods until they reach the end consumer.

See Fig. 8.3 symbolically related to product distribution networks.

8.3 Basics of Spatial Grasp Model and Technology

Within Spatial Grasp Model and Technology (SGT), described in Chaps. 3, 4 and 5, also in [75–92] and many other papers, a high-level operational scenario expressed in recursive Spatial Grasp Language (SGL), starting in any world point or points, *propagates, covers, and matches the distributed environment in parallel wavelike mode*, as symbolically shown in Fig. 8.4. Such propagation can result in returning and analyzing the reached states and data, which may be arbitrarily remote, or used for launching more waves.

The SGL allows for organizing *direct space presence and operations* with unlimited powers and parallelism. Its universal recursive organization, with operational

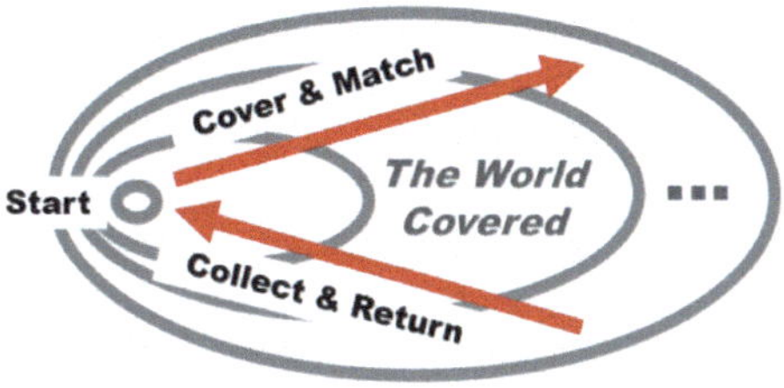

Fig. 8.4 Parallel wavelike world coverage under spatial grasp model

scenarios called *grasp*, can be expressed by just a single formula:

$$\textbf{\textit{grasp}} \rightarrow \textbf{\textit{constant}} \mid \textbf{\textit{variable}} \mid \textbf{\textit{rule}} \; (\{\textbf{\textit{grasp}},\})$$

The *rule* expresses certain action, control, description or context accompanied with operands, which can be any *grasps* too. The rules, starting in certain points, can organize navigation of the world sequentially, in parallel, or any combinations thereof. They can result in the same application points or cause movement to other world points with obtained results left there or returned.

The *SGL interpreter* consists of specialized modules serving SGL scenarios or their parts happened to be inside this interpreter, also organizing exchanges with other interpreters for distributed SGL scenarios. Communicating SGL interpreters can be in arbitrary number of copies, representing *powerful spatial engines operating without central resources or control*, as in Fig. 8.5. They can effectively work with *spatial graph and network data* of any volumes and distributions. Details of SGT are explained in Chaps. 3, 4 and 5.

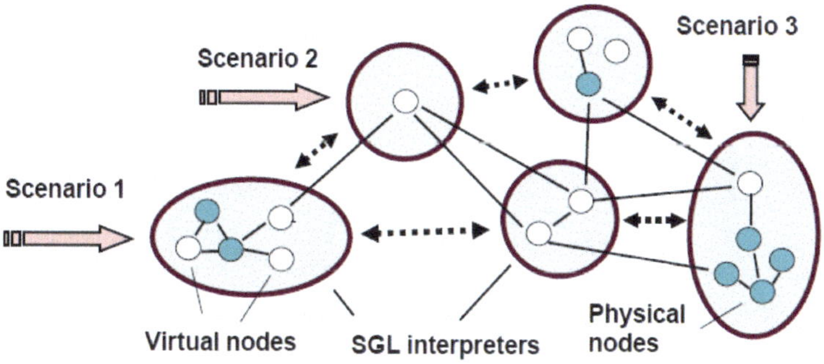

Fig. 8.5 SGL distributed networked interpretation

8.4 Practical Examples of Combined Networking Solutions

8.4.1 Product Development Networks

Some basic features of production networks, with sources [93, 94] reviewed in Sect. 8.2, include the following. A global production network (GPN) has interconnected nodes and links which extend spatially across national and international boundaries. Nowadays, companies of any size operate globally in the form of GPNs, where globalization has two sides: (a) lack of geographical barriers, increased mobility, flux of capital, goods and ideas; and (b) necessity for concentration.

In Fig. 8.6, three symbolic production networks are shown developing products of types A, B, and C (with links inside them named with corresponding small letters a, b, c) and having nodes of different shapes (like square, hexagon, circle), also painted in different colors (their connecting links too). Possible business links between nodes of different networks are mentioned by broken black x lines.

The importance of different interlinked production components in these networks can be analyzed by the network structures in a variety of ways, and first of all, using *node centrality* measures like those discussed in [70] and Chap. 7, named as: *degree centrality*, *closeness centrality*, *betweenness centrality*, and *eigenvector centrality*, with some of them considered below.

In ***Degree Centrality***, the degree of a node is defined as the number of direct connections a node has with other nodes. The following SGL scenario, applied in parallel in all network nodes of the chosen type (let it be A in Fig. 8.6), will result with node names accompanied by heir degrees (number of nodes directly connected with them by type a links for this case).

```
output(hop(all_nodes(TYPE == A));
    append(count(hop(all_links(a))), NAME))
```

Result will be: 2:A1, 2:A2, 5:A3, 2:A4, 3:A5.

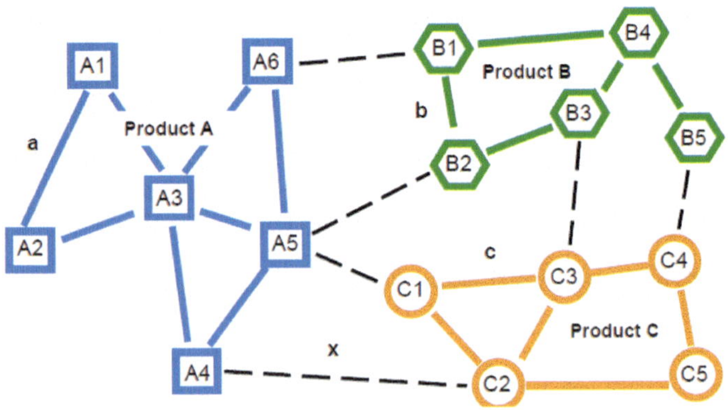

Fig. 8.6 Production network

To name the node of type A having maximum degree, which the use of links type a, we may write:

```
output_max(hop(all_nodes(TYPE == A));
      append(count(hop(all_links(a))), NAME))
```

Will be having: 5:A3.

But if we include nodes of type A also connected to other type nodes (i.e. by links x of Fig. 8.6).

```
output_max(hop(all_nodes(TYPE == A));
      append(count(hop(all_links(any))), NAME)
```

will have either 5:A3 or 5:A5.

For ***Closeness Centrality***, distances to all other nodes from each node and their summary can be found by the following SGL scenario also naming the node having minimum of such sums (i.e. being the *most central*).

```
output_min(
 hop(all_nodes(TYPE == A)); IDENTITY = NAME;
 frontal(Far) = 0; nodal(Distance);
 stay(repeat(hop(hop(all_links(a)); Far += 1;
       or(Distance == nil, Distance > Far);
       Distance = Far));
 append(sum(hop(all_other_nodes(TYPE == A));
     Distance), NAME))
```

Result will be 5:A3, which happened to be the same as for the degree centrality.

But if to consider all links related to nodes of type A, as of Fig. 8.6 (i.e. named a and x), after making the following change in the previous scenario:

```
hop(all_links(a)) → hop(all_links(any)),
```

we will receive 9:A5, whereas A3 will result with 10:A3.

We can similarly examine parameters of the networks of types B and C shown in Fig. 8.6. Other centrality measures may also be used for such networks as of Fig. 8.6, like *betweenness centrality* and *eigenvector centrality*, described in detail in [70].

8.4.2 Product Assessment and Delivery Request

Some basic features of product assessment and request networks, with sources [95–97] were investigated in Sect. 8.2. Evaluation of product development systems by network analysis has become an increasingly crucial aspect of product design. Network algorithms are used for expressing product opinions, helping consumers evaluate and compare products and services. The product assessment models provide methods to evaluate product design and empower development process.

In Fig. 8.7 we are symbolically showing sets of interlinked users ui which are interested in particular products developed by networks of types A, B, and C, as of

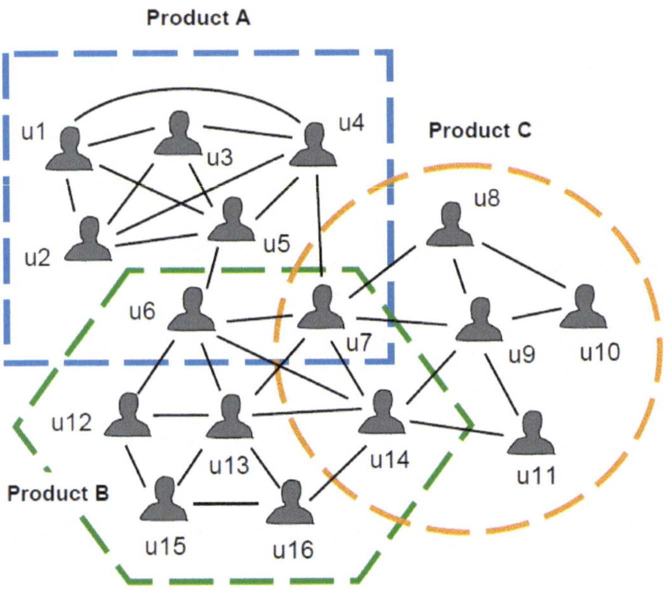

Fig. 8.7 Product assessment networks

Fig. 8.6, having different colors and shapes of their nodes (i.e. squares, hexagons, and circles). In Fig. 8.7 these shapes are used as borders of regions covering users having interest in these particular products (same users may fall under more than a single product and region influence). There may be different methods of how to discover user communities most interested in a particular project, for example, based on how intensely they interlink and communicate with each other in relation to this product.

The strongest groupings of graph nodes are classically considered as cliques, which we will be trying to discover in relation to product A, analyzing the square blue region of Fig. 8.7 (its development network was investigated in the previous session). We are providing below a universal solution in SGL for finding customer cliques related to product A (i.e. maximum full sub-graphs in the customer interaction network), assuming their number of nodes exceeding some threshold to represent a realistic interest.

```
hop(all_nodes(customer, includes(INTEREST, A));
frontal(Clique) = NAME;
repeat(
 hop(links(all),
   nodes(customer, includes(INTEREST, A)));
 not_belong(NAME, Clique);
 yes(and_parallel(hop(links_any, nodes(Clique))));
 if(PREDECESSOR> NAME, append(Clique, NAME), blind));
 count(Clique) >= 4;
 output(hop(Clique); append(NAME, DEMAND))
```

This scenario, starting in all nodes and following their links to other nodes in parallel, is collecting node names in new individual hops which have links with all previously collected nodes unless such nodes cannot be found. The finally collected set of node names in the frontal variable `Clique` is followed by parallel hop to all clique nodes with issuing their names together with related product value demands, the latter picked up by special nodal variable `DEMAND` in each node. As same cliques can be grown starting from all their any nodes and evolved in any order by moving through other nodes, this duplicity is easily removed by allowing the clique nodes to proceed only in the reduced value of their names, which is unique (this can also be organized in the opposite way).

Reply for the obtained five-node clique of Fig. 8.7 related to product `A`, with node names and discovered product demands will be:

(u1:7, u2:8, u3:2, u4:9, u5:5).

8.4.3 Product Delivery to Chosen Customers

Some basic features of product delivery networks, with sources [98–100] reviewed in Sect. 8.2, include the following. The distribution networks have to interconnect the consumption and production sides of the supply chain. The network consists of storage facilities, warehouses, and transportation systems that support movement of goods until they reach the end consumer. The value delivery network emphasizes the importance of understanding and meeting customer needs by providing high delivered value.

Imagine we already have a distributed physical network (including road, see, air and possibly other lines) using which we want to deliver a product from the production network, say, from node `A3` of Fig. 8.6, which showed the highest value by both degree and closeness centrality measures, to the interested in this product physically existing consumer nodes. The latter may represent the chosen clique nodes `u1-u5` in the assessment network in Fig. 8.7, together with the discovered product `A` current volume demands. We may first find the shortest path tree of this network originating from `A3` node and leading to all other nodes (as in Fig. 8.8a), and then leave only paths in it from `A3` to all `u1-u5` nodes, altogether forming the *best delivery infrastructure* to particular users for the concrete product (as in Fig. 8.8b, where some consumer nodes may happen to be on the shortest paths to other consumer nodes).

The nodes of this (Fig. 8.8b) infrastructure are shown with two associated numerical values, with the second reflecting the value of product they requested personally, and the first one as amount of product needed to reach them, which represents the sum of their own need and the rest to be delivered to other nodes via them. The related delivery scenario may be as follows. It starts in node `A3` with the total product volume needed by all customers which is then gradually split and reduced during physically moving down through the delivery infrastructure to the defined users, ultimately leaving with the latter only the amount requested by them. For the physical product delivery process, special nodal variable `Store` is used in each node, containing

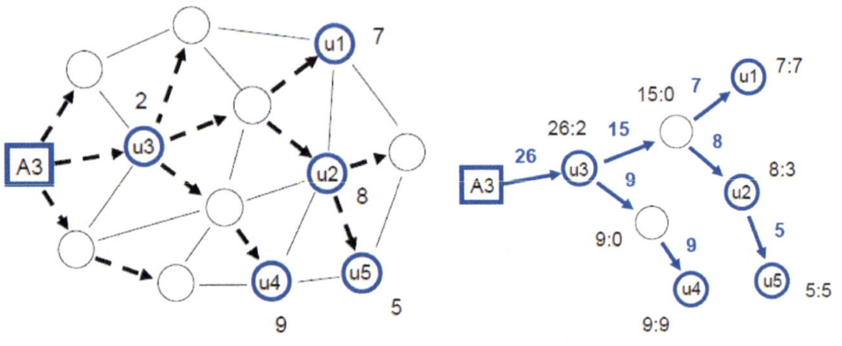

Fig. 8.8 Forming optimal delivery infrastructure to particular users

at `Start` (i.e., `A3`) the product total volume, and then reducing, if needed, in the following shortest path delivery nodes. At the end of distribution, `Store` will contain only the volume individually requested by each node.

```
nodal(Dist, Up, Up2,
   Fin = (u1:7, u2:8, u3:2, u4:9, u5:5));
frontal(Far, Demand, Pack); nodal(Start = A3, Full,
    Store);
hop(Start);
sequence(
 repeat(
  hop_links(all); Far += LENGTH;
  or(Dist == nil, Dist > Far);
  Dist = Far; Up = BACK),
 (wait(split(Fin); Demand = VAL[2]; hop(VAL[1]);
  Full = Demand);
  repeat(Up2 = Up; hop(Up);
      Full += Demand; Demand = Full)),
 (Store = assemble(product, Full);
  repeat(hop_links(all); Up2 == BACK;
      Demand = Full; hop(BACK);
      Pack = withdraw(Store, Demand);
      move(BACK); Store = Pack))
```

8.5 Summary of the Combined Networking Solutions

In the previous subsections we showed separate but logically interconnected solutions of solving rather complex integral problem of the networked based creation of products, their assessment by customers, and then physical delivery of the chosen products to the most interested customers, selected too. All this combined multi-network heterogeneous process, with same nodes belonging to very different networks, may

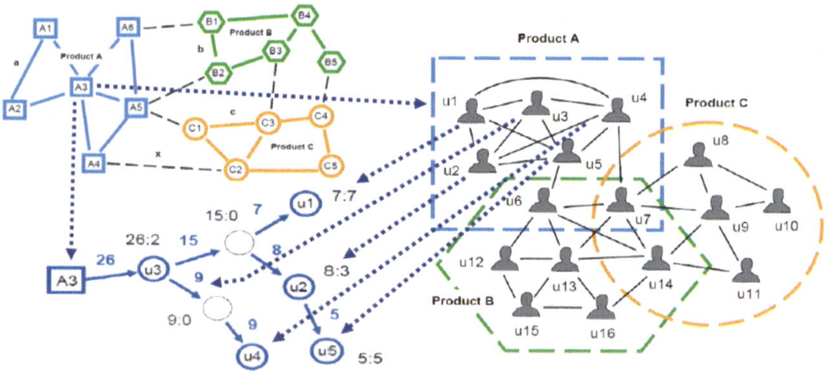

Fig. 8.9 Summary of combined multi-networking solution

be exhibited as the united one, symbolically shown Fig. 8.9. We can easily integrate in SGL all these and much more complex internetworking representations and solutions into the global production-assessment-distribution social infrastructure. In it, the transference of parameters between heterogeneous networks (shown by dotted arrows in Fig. 8.9), can be done naturally and automatically within the social global integrity, awareness and consciousness process, as discussed in [78, 79, 81, 84, 87].

In the above examples we showed possible practical solutions related to product A and its production network, like in Fig. 8.9. But we can similarly organize these for other product types B and C, their networks, and communities of interested in them customers, also for much more complex united products with their collective development by different types of production networks.

8.6 Conclusions

The results of this chapter confirm availability and efficiency of using SGT and SGL for investigating, expressing, and implementing most basic network and graph creation and management issues. SGL allows us to provide detailed, very clear and compact solutions for the network problems which can operate on arbitrary large and complex networks, and in highly parallel and fully distributed mode. Practical cases were investigated on integration of different types of networks within global production-assessment-distribution social infrastructures providing solutions potentially unachievable by separate networks or their mere collections. Deep integration of heterogeneous distributed systems can be organized naturally and automatically within the global integrity, awareness, and consciousness processes under Spatial Grasp paradigm, as also shown and discussed in previous publications.

References

1. Transport networks. Inspire Registry (2015). https://inspire.ec.europa.eu/theme/tn
2. Transportation networks. Travel Forecasting Resource. https://tfresource.org/topics/Transportation_networks.html
3. Durkin, J., Goodman, J. et al. Building Multiservice Transport Networks (Networking Technology), 1st edn. Cisco Press (2012, January 10)
4. Murakami, M., Li, H., Ryoo, J-D.: Packet Transport Networks: Overview and Future Direction. In: 2nd APT/ITU Conformance and Interoperability Workshop (C&I-2) (2014, August). https://www.itu.int/en/ITU-T/C-I/interop/Documents/20140826/CI-2_INP-09_Packet_Transport_Networks_Overview_and_Future_Direction.pdf
5. Air Transport Network. https://en.wikipedia.org/wiki/Air_Transport_Network
6. Qureshi, K.N., Ul Islam, F. et al.: Improved road segment-based geographical routing protocol for vehicular ad-hoc networks. Electronics **9**(8), 1248 (2020). https://doi.org/10.3390/electronics9081248. https://www.mdpi.com/2079-9292/9/8/1248
7. Sujan: What is communication network? Definition, Importance, and 4 Types (Explained) (2023, June 2). https://tyonote.com/communication_network/
8. Fitzek, F.H.P., Granelli, F. et al.: Computing in Communication Networks: From Theory to Practice, 1st edn. Academic Press (2020, June 4)
9. Srikant, R., Ying, L.: Communication Networks: An Optimization, Control, and Stochastic Networks Perspective, 1st edn. Cambridge University Press (2013, November 14)
10. Raghunandan, K.: Introduction to Wireless Communications and Networks: A Practical Perspective (Textbooks in Telecommunication Engineering), 1st ed. 2022 edn. Springer (2023, April 2)
11. Lou, Y., Ahmed, N.: Underwater Communications and Networks (Textbooks in Telecommunication Engineering). Springer (2021, November 15)
12. Kaufman, C., Perlman, R. et al.: Network Security: Private Communication in a Public World (Prentice Hall Series in Computer Networking and Distributed Systems), 3rd edn. Addison-Wesley Professional (2022, September 26)
13. Wu, Y., Singh, S. et al.: 6G Mobile Wireless Networks (Computer Communications and Networks), 1st ed. 2021 edn. Springer (2022, August 26)
14. Social network. https://en.wikipedia.org/wiki/Social_network
15. Kenton, W.: What is social networking? Investopedia. https://www.investopedia.com/terms/s/social-networking.asp#:~:text=Social%20networking%20refers%20to%20using,)%2C%20Instagram%2C%20and%20Pinterest. Accessed 24 May 2023
16. Nanawati, S.: Social Network Analytics. Published in Analytics Vidhya (2019, August 1). https://medium.com/analytics-vidhya/social-network-analytics-f082f4e21b16
17. Singh, C.: 20+ Most Popular Social Networking Sites in 2024 (2023, December 17). https://www.socialpilot.co/social-networking-sites
18. Jackson, M.O.: Social and Economic Networks, Illustrated edn. Princeton University Press (2010, November 21)
19. Harrison, T.: Battle Networks and the Future Force. Center for Strategic and International Studies (2021). https://www.csis.org/analysis/battle-networks-and-future-force
20. Stillion, J., Clark, B.: What It Takes to Win: Succeeding in 21st Century Battle Network Competitions. Center for Strategic and Budgetary Assessments (CSBA) (2015). https://csbaonline.org/uploads/documents/What-it-Takes-to-Win.pdf
21. Hadley, G.: DAF outlines a new 'Battle Network' as its contribution to JADC2. Air & Space Forces Magazine (2023, March 23). https://www.airandspaceforces.com/daf-battle-network-contribution-jadc2/
22. The Challenges of Military Networks and Technological Developments. Elbit Systems. https://elbitsystems.com/e-lynx_multi_channel/
23. Themnér, A., Karlén, N.: Building a safety net: explaining the strength of ex-military networks. Secur. Stud. **29**(2), 268–300 (2020). https://doi.org/10.1080/09636412.2020.1722851

24. What is an economic network? Advantages and disadvantages of economic networks. Vietnambiz (2019). https://vietnambiz.vn/mang-luoi-kinh-te-economic-network-la-gi-uu-nhuoc-diem-cua-mang-luoi-kinh-te-20191011113428212.htm
25. Halton, C.: Economic network: what it is. Pros and Cons, Examples. https://www.investopedia.com/terms/e/economic-network.asp#:~:text=Key%20Takeaways-,An%20economic%20network%20is%20a%20combination%20of%20individuals%2C%20groups%2C%20or,companies%20or%20partnerships%20between%20corporations. Accessed 24 October 2022
26. Todo, Y., Matous, P.: Economic networks for more innovative and resilient economies (2015, June 26). https://cepr.org/voxeu/columns/economic-networks-more-innovative-and-resilient-economies
27. Emmert-Streib, F., Tripathi, S. et al.: Understanding the world economy in terms of networks: a survey of data-based network science approaches on economic networks. Front. Appl. Math. Stat. (Sec. Mathematical Finance) **4** (2018). https://doi.org/10.3389/fams.2018.00037
28. Wixted, B.: Innovation System Frontiers: Cluster Networks and Global Value (Advances in Spatial Science). Springer (2009)
29. Manyika, J.: What the economy looks like so far for individuals in the 21st century (2020, February 6). https://www.linkedin.com/pulse/what-economy-looks-like-so-far-individuals-21st-century-james-manyika
30. Hayes, A.: Distribution network: definition, how it works, and examples. Investopedia. https://www.investopedia.com/terms/d/distribution-network.asp#:~:text=In%20a%20supply%20chain%2C%20a,or%20through%20a%20retail%20network. Accessed 22 Feb 2021
31. Distribution Network: The flow of goods from a producer or supplier to an end consumer. Written by CFI Team. https://corporatefinanceinstitute.com/resources/valuation/distribution-network/
32. What are types of distribution networks in supply chain management? oboloo FAQ's. https://oboloo.com/blog/what-are-types-of-distribution-networks-in-supply-chain-management/
33. Distribution network—what is it and what are the benefits? https://www.salesbabu.com/blog/what-is-a-distribution-network-and-its-benefits/
34. Wu, Q., Shen, F. et al.: Optimal Operation of Active Distribution Networks: Congestion Management, Voltage Control and Service Restoration. Academic Press (2023)
35. Watts, S.: What is a Virtual Network? BMC (2020, June 16). https://www.bmc.com/blogs/virtual-network/
36. What is Virtual Networking? VMware by Broadcom. https://www.vmware.com/topics/glossary/content/virtual-networking.html
37. Network virtualization. https://en.wikipedia.org/wiki/Network_virtualization
38. Carlos, O., Duarte, M.B., Pujolle, G.: Virtual Networks: Pluralistic Approach for the Next Generation of Internet (Networks and Telecommunications), 1st edn. Wiley-ISTE (2013, August 2)
39. Fundamentals of Virtual Networking. https://www.geeksforgeeks.org/fundamentals-of-virtual-networking/
40. Neural network. https://en.wikipedia.org/wiki/Neural_network
41. Anand, V.: Introduction to Neural Network in Machine Learning. https://www.analyticsvidhya.com/blog/2022/01/introduction-to-neural-networks/. Accessed 10 August 2023
42. Ashtari, H.: What is a Neural Network? Definition, Working, Types, and Applications in 2022. https://www.spiceworks.com/tech/artificial-intelligence/articles/what-is-a-neural-network/. Accessed 3 August 2022
43. Artificial Neural Network. NVIDIA Developer. https://developer.nvidia.com/discover/artificial-neural-network
44. Wu, L., Cui, P. et al.: Graph Neural Networks: Foundations, Frontiers, and Applications, 1st ed. 2022 edn. Springer (2022, January 4)
45. Borsboom, D., Deserno, M.K., Rhemtulla, M. et al.: Network analysis of multivariate data in psychological science. Nat. Rev. Methods Prim. **1**, 58 (2021). https://doi.org/10.1038/s43586-021-00055-w. https://www.nature.com/articles/s43586-021-00055-w

46. Jones, P.J., Mair, P., McNally, R.J.: Visualizing psychological networks: a tutorial in R Front. Psychol. (Sec. Quantitative Psychology and Measurement) **9** (2018). https://doi.org/10.3389/fpsyg.2018.01742
47. Fonseca-Pedrero, E.: Network analysis in psychology. Psychol. Pap. **39**(1), 1–12 (2018). https://doi.org/10.23923/pap.psicol2018.2852. https://www.papelesdelpsicologo.es/pdf/2852.pdf
48. Bringmann, L.F.: Person-specific networks in psychopathology: past, present, and future. Curr. Opin. Psychol. **41**, 59–64 (2021). https://www.sciencedirect.com/science/article/pii/S2352250X21000300?via%3Dihub
49. Yuqing, C., Shuyang, D. et al.: Network analysis and its applications in psychology. Adv. Psychol. Sci. **28**(1), 178–190 (2020). https://doi.org/10.3724/SP.J.1042.2020.00178
50. Campana, P.: Explaining criminal networks: strategies and potential pitfalls. Methodol. Innov. **9**, 1–10 (2016). https://doi.org/10.1177/2059799115622748
51. Bichler, G.: Understanding Criminal Networks: A Research Guide, 1st edn. University of California Press (2019, September 24)
52. Morselli, C.: Crime and Networks (Criminology and Justice Studies). Routledge (2013)
53. New forms of organized crime: networked structure. UNODS, Unite Nations Office on Drugs and Crime. https://www.unodc.org/e4j/en/organized-crime/module-7/key-issues/networked-structure.html
54. Hufnagel, S., Moiseienko, A.: Criminal Networks and Law Enforcement: Global Perspectives On Illegal Enterprise (Transnational Criminal Justice), 1st edn. Routledge (2019, June 27)
55. Detecting and disrupting transnational criminal organizations: analytics for interdependent smuggling and money-laundering networks. Criminal Investigation and Network Analysis. DHS Centers of Excellence. https://cina.gmu.edu/projects/detecting-and-disrupting-transnational-criminal-organizations-analytics-for-interdependent-smuggling-and-money-laundering-networks/
56. An Introduction to Networks. https://mathinsight.org/network_introduction
57. The Degree Distribution of a Network. https://mathinsight.org/degree_distribution
58. Random Networks. https://mathinsight.org/random_network
59. Network Algorithms. https://ics.uci.edu/~goodrich/teach/cs165/notes/NetworkAlgs.pdf
60. Graph and Network Algorithms. https://ch.mathworks.com/help/matlab/graph-and-network-algorithms.html
61. Schmid, S., Mandal, P.S.: Distributed Network Algorithms (Lecture Notes for GIAN Course) (2016 Summer). https://schmiste.github.io/GIAN-Lecture-Notes-NetAlg.pdf
62. Tarjan, R.E.: Data structures and network algorithms. In: CBMS-NSF Regional Conference Series in Applied Mathematics. https://doi.org/10.1137/1.9781611970265
63. Ravikiran, A.S.: What is Graph in Data Structure & Types of Graph? (2023, July 26). https://www.simplilearn.com/tutorials/data-structure-tutorial/graphs-in-data-structure
64. Graph Operations. https://pages.cs.wisc.edu/~paton/readings/Old/fall01/GRAPH-OPS.html
65. Basic Graph Theory. https://learn.saylor.org/mod/page/view.php?id=27304
66. Pandey, K.: Graph Data Structure—Explained With Examples (2022, July 5). https://www.masaischool.com/blog/graph-data-structure-explained-with-examples/
67. Ford-Fulkerson Algorithm for Maximum Flow Problem. https://www.scaler.com/topics/data-structures/ford-fulkerson-algorithm-for-maximum-flow-problem/
68. Bhasin, J.: Graph Analytics—Introduction and Concepts of Centrality, Towards Data Science (2019, Aug 14). https://towardsdatascience.com/graph-analytics-introduction-and-concepts-of-centrality-8f5543b55de3
69. Rodrigues, F.A.: Network centrality: an introduction (2019, January 22). arXiv preprint arXiv: 1901.07901v1. https://arxiv.org/pdf/1901.07901.pdf
70. Disney, A.: Social network analysis 101: centrality measures explained. Cambridge Intelligence (2020, January). https://cambridge-intelligence.com/keylines-faqs-social-network-analysis/
71. Mussone, L., Viseh, H., Notari, R.: Novel centrality measures and applications to underground networks. Phys. A: Stat. Mech. Appl. **589**, 126595 (2022) https://www.sciencedirect.com/science/article/abs/pii/S0378437121008633

72. Bringmann, L.F., Elmer, T., Epskamp, S., Krause, R.W. et al.: What do centrality measures measure in psychological networks? J. Abnorm. Psychol. **128**(8), 892–903 (2019). https://doi.org/10.1037/abn0000446. 0021-843X/19/$12.00. https://www.apa.org/pubs/journals/features/abn-abn0000446.pdf
73. Meghanathan, N.: Centrality Metrics for Complex Network Analysis: Emerging Research and Opportunities (Advances in Wireless Technologies and Telecommunication). IGI Global (2018). https://www.igi-global.com/book/centrality-metrics-complex-network-analysis/181520#table-of-contents
74. Lange, A.: Centrality in Strategic Transportation Network Design: An Application to Less-Than-Truckload Networks (Edition KWV). Global Production Network, Springer Gabler (2019). https://en.wikipedia.org/wiki/Global_production_network
75. Sapaty, P.S.: A distributed processing system. European Patent N 0389655, Publ. 10.11.93, European Patent Office
76. Sapaty, P.S.: Mobile Processing in Distributed and Open Environments. John Wiley & Sons, New York (1999)
77. Sapaty, P.S.: Ruling Distributed Dynamic Worlds. John Wiley & Sons, New York (2005)
78. Sapaty, P.S.: Managing Distributed Dynamic Systems with Spatial Grasp Technology. Springer (2017)
79. Sapaty, P.S.: Holistic Analysis and Management of Distributed Social Systems. Springer (2018)
80. Sapaty, P.S.: Complexity in International Security: A Holistic Spatial Approach. Emerald Publishing (2019)
81. Sapaty, P. S.: Symbiosis of Real and Simulated Worlds under Spatial Grasp Technology. Springer (2021)
82. Sapaty, P.S.: Spatial Grasp as a Model for Space-based Control and Management Systems. CRC Press (2022)
83. Sapaty, P.S.: The Spatial Grasp Model: Applications and Investigations of Distributed Dynamic Worlds. Emerald Publishing (2023)
84. Sapaty, P.S.: Providing Integrity, Awareness, and Consciousness in Distributed Dynamic Systems. CRC Press (2025)
85. Sapaty, P.S.: Spatial management of air and missile defence operations. Math. Mach. Syst. 1 (2023). http://www.immsp.kiev.ua/publications/articles/2023/2023_1/01_23_Sapaty.pdf
86. Sapaty, P.S.: Providing Distributed System Integrity under Spatial Grasp Technology. Math. Mach. Syst. 2 (2023). http://www.immsp.kiev.ua/publications/articles/2023/2023_2/02_23_Sapaty.pdf
87. Sapaty, P.S.: Providing global awareness in distributed dynamic systems. Int. Relat. Dipl. **11**(2), 87–100 (2023). https://doi.org/10.17265/2328-2134/2023.02.002. https://www.davidpublisher.com/Public/uploads/Contribute/6486c3d05a6cc.pdf
88. Sapaty, P.S.: Simulating distributed consciousness with spatial grasp model. Math. Mach. Syst. 3 (2023)
89. Sapaty, P.S.: Network centrality operations under spatial grasp technology. J. Adv. Artif. Intell. Mach. Learn. **1**(1), 1–11 (2023). https://www.scivisionpub.com/journals/articleinpress-journal-of-advances-in-artificial-intelligence-and-machine-learning
90. Sapaty, P.S.: Managing distributed systems with spatial grasp patterns. Math. Mach. Syst. 4 (2023)
91. Sapaty, P.S.: Spatial networks as models for organoid cultures and brain research. Math. Mach. Syst. 1 (2024)
92. Sapaty, P.S.: Basic network operations under spatial grasp model. Sci. Set J. Phys. **3**(1), 01–16 (2024)
93. Lanza, G., Ferdows, K., Kara, S. et al.: Global production networks: design and operation. CIRP Ann. **68**(2) (2019). https://www.sciencedirect.com/science/article/abs/pii/S0007850619301659
94. Gonzaga, F.P.: Local centers and connectivity: the foundation of global production networks (1919). https://www.linkedin.com/pulse/local-centers-connectivity-foundation-global-networks-filipe

95. Collins, S., Yassine, A.A., Borgatti, S.P.: Evaluating product development systems using network analysis. Syst. Eng. **12**(1), 55–68 (2009) https://doi.org/10.1002/sys.20108. https://www.researchgate.net/publication/227602841_Evaluating_Product_Development_systems_using_network_analysis
96. Lin, H., Deng, X. et al.: A study of sustainable product design evaluation based on the analytic hierarchy process and deep residual networks. Sustainability **15**(19), 14538 (2023) https://doi.org/10.3390/su151914538. https://www.mdpi.com/2071-1050/15/19/14538
97. Danilchenko, K., Segal, M., Vilenchik, D.: Opinion Spam Detection: A New Approach Using Machine Learning and Network-Based Algorithms. arXiv preprint arXiv:2205.13422v1 [cs.LG] (2022, May 26). https://www.semanticscholar.org/reader/1a6b30f29e2c9eee9eb34586bd9a1f9e822a5962
98. Bajor, P., Horvath, A.: The role of decision-making parameters in constructing and re-engineering of distribution networks. https://www.researchgate.net/publication/23637165_The_role_of_decision-making_parameters_in_constructing_and_re-engineering_of_distribution_networks
99. What is value delivery network in marketing? SCISPACE. https://typeset.io/questions/what-is-value-delivery-network-in-marketing-4um9b4s3z1
100. Distribution Network: The flow of goods from a producer or supplier to an end consumer. CFI Team. https://corporatefinanceinstitute.com/resources/valuation/distribution-network/

Chapter 9
Active Spatial Patterns in SGL Versus Traditional Algorithms

9.1 Introduction

The pattern is everything in the digital world. Patterns can be seen physically or observed mathematically by applying algorithms. A pattern, for example, can be as follows:

- An arrangement of lines or shapes
- Regularity in the world
- Human-made design or abstract ideas
- Unvarying way of acting or doing
- A model, plan, or diagram used as a guide
- A standard way of moving, acting, etc.
- Representation of a class or type
- An example, sample, instance, or specimen
- A distinctive style, model, or form
- A combination of qualities, acts, tendencies, etc.
- An original or model deserving imitation.

The aim of this chapter is to investigate and analyze the applicability of the developed Spatial Grasp Model and Technology and especially its basic Spatial Grasp Language, which were presented in detail in Chaps. 3, 4 and 5, for the representation and implementation of different types of patterns that can be used in the simulation and management of a variety of distributed dynamic systems. These systems may cover such areas as education, economy, science, ecology, psychology, security, defense, international relations, space research, and many others [1–55], as investigated in Chap. 2.

The rest of the chapter is organized as follows. Section 9.2 reviews existing works on patterns including: pattern definition and theory, different pattern types, pattern matching and recognition, and pattern languages. Section 9.3 briefs Spatial Grasp Model and Technology described in detail in Chaps. 3, 4 and 5, also in [56–73] and

P. S. Sapaty, *Spatial Networking in the United Physical, Virtual, and Mental World*, Studies in Systems, Decision and Control 533,
https://doi.org/10.1007/978-3-031-62154-3_9

other publications. Section 9.4 considers creating specific patterns in SGL like regular patterns and patterns of concrete objects. Section 9.5 offers pattern-based practical solutions for managing transport column, collecting zone coordinates, and tracking mobile objects. Section 9.6 offers pattern recognition examples in SGL. Section 9.7 provides pattern matching solutions for arbitrary and tree-like graph patterns with their different variants. Section 9.8 explains how SGL can be considered as a real pattern language. Section 9.9 concludes the chapter with confirmed effectiveness of SGL and SGT for managing patterns in different areas. The References contain analyzed networks in different areas, reviewed pattern-based sources, and published papers and books on SGT and SGL.

9.2 Review of Existing Works on Patterns

After being searched for, analyzed, compared, discussed and classified, the discovered pattern-based ideas, sources and publications can be grouped as follows.

- **Pattern definition and theory**

The pattern [74] is regularity in the world, human-made design, or abstract ideas. As such, the pattern elements repeat in a predictable manner. A geometric pattern is a kind of pattern formed of geometric shapes and typically repeated like a wallpaper design. Any of the senses may directly observe patterns. Conversely, abstract patterns in science, mathematics, or language may be observable only through analysis. See some pattern examples in Fig. 9.1.

Pattern Theory: From Representation to Inference [75] provides a comprehensive and accessible overview of the modern challenges in signal, data, and pattern analysis in speech recognition, computational linguistics, image analysis, and computer vision. It is aimed at graduate students in biomedical engineering, mathematics, computer science, and electrical engineering with a good background in mathematics and probability.

Pattern Theory-based Interpretation of Activities [76] presents a novel framework based on Grenander's pattern theoretic concepts for high-level interpretation of video activities. This framework allows us to elegantly integrate ontological constraints and machine learning classifiers in a single formalism to construct high-level semantic interpretations that describe video activity.

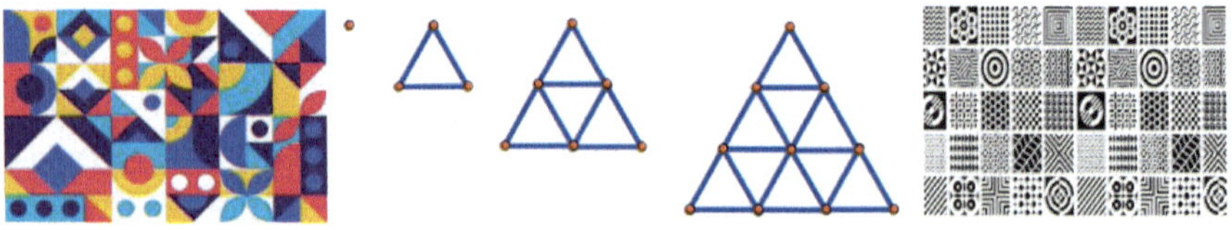

Fig. 9.1 Geometric pattern examples

Pattern Theory: A Unifying Perspective [77] introduced the term "pattern theory" as a name for a field of applied mathematics that gave a theoretical framework for a large number of related ideas, techniques, and results from such fields as computer vision, speech recognition, image and acoustic signal processing, pattern recognition, neural nets, and parts of artificial intelligence. Pattern theory contains the foundational elements of a universal theory of thought itself, one of which stands in opposition to the accepted analysis of thought in terms of logic.

Pattern theory [78] is a mathematical formalism used to describe the knowledge of the world as patterns. It differs from other approaches to artificial intelligence in that it does not begin with assigning algorithms and machinery to recognize and classify patterns, it uses a vocabulary to articulate and recast the pattern concepts in precise language. Pattern theory spans algebra and statistics, as well as local topological and global entropic properties.

- **Different pattern types**

Patterns of distributed systems are discussed in [79]. Distributed systems provide a particular challenge to a program. Despite this, many organizations rely on a range of core distributed software for handling data storage, messaging, system management, and computing capability. These systems face common problems which they solve using similar solutions. The paper recognizes and develops these solutions as patterns that help to understand, communicate, and teach distributed system design.

Statistical descriptions of spatial patterns are discussed in [80]. Spatial statistics can be defined as a statistical description of spatial data and a spatial pattern or process. Spatial statistics allow a quantitative description along with indications of statistical significance in observational data on a pattern or a process operating in space. This quantitative and statistical description allows the exploration and modeling of spatial patterns and processes and their relationships with other spatial phenomena.

Recursive pattern: a technique for visualizing very large amounts of data is considered in [81]. An important goal of visualization technology is to support the exploration and analysis of very large amounts of data. This paper proposes a new visualization technique called a «recursive pattern» which has been developed for visualizing large amounts of multidimensional data. The technique is based on a generic recursive scheme that generalizes a wide range of pixel-oriented arrangements for displaying large data sets.

The focus on the hidden pattern: a patternist philosophy of mind is made in [82]. The Hidden Pattern presents a novel philosophy of mind, intended to form a coherent conceptual framework within which it is possible to understand the diverse aspects of mind and intelligence in a unified way. The central concept of the philosophy presented is the concept of "pattern": minds and the world they live in and co-create are viewed as patterned systems of patterns, evolving over time.

Workflow control-flow patterns are studied in [83]. The Workflow Patterns Initiative aims at delineating fundamental requirements that arise during business process modeling and describing them in an imperative way. The first deliverable of this

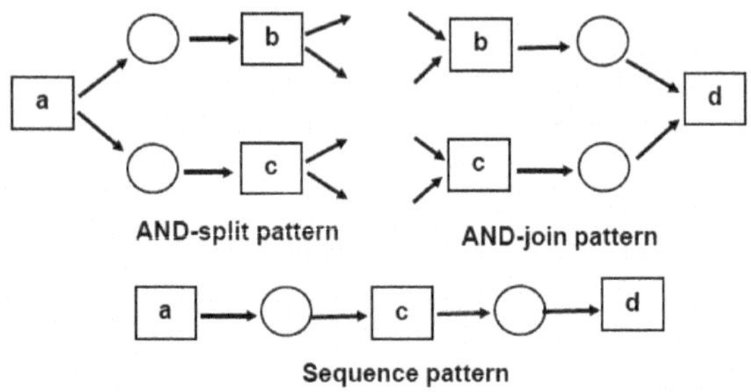

Fig. 9.2 Examples of workflow patterns

project was a set of twenty patterns describing the control-flow perspective of workflow systems. These patterns have been widely used by practitioners, vendors, and academics in the selection, design, and development of workflow systems.

Workflow patterns are also considered in [84]. Requirements for workflow languages are indicated through workflow patterns. In this context, patterns address business requirements in an imperative workflow style expression. The paper describes a number of workflow patterns that can identify comprehensive workflow functionality. These patterns provide a basic comparison of a number of commercially available workflow management systems. See related pattern examples in Fig. 9.2.

- **Pattern matching and recognition**

Pattern matching is discussed in [85]. In computer science, pattern matching is the act of checking a given sequence of tokens for the presence of the constituents of any pattern. In contrast to pattern recognition, the match usually has to be exact: either it will or will not be a match. The uses of pattern matching include outputting the locations of a pattern within a token sequence to output some component of the matched pattern and to substitute the matching pattern with another token.

Fast graph pattern matching is studied in [86]. Due to the rapid growth of Internet technology and new scientific and technological advances, the number of applications that model data as graphs increases because graphs have high expressive power to model complicated structures. The dominance of graphs in real-world applications requires new graph data management so that users can access graph data effectively and efficiently. This paper studies a graph pattern-matching problem over a large data graph.

Pattern matching in massive metadata graphs at scale is presented in [87]. Pattern matching in graphs, that is finding sub-graphs that match a smaller template graph within the large background graph, is fundamental to graph analysis and serves a wide range of applications. Existing solutions have limited scalability, are difficult to parallelize, support only a limited set of search patterns, and/or focus on only a subset of the real-world problems. This work explores ways of designing a scalable solution for subgraph pattern matching. See the related example in Fig. 9.3.

Pattern recognition [88] is the automated recognition of patterns and regularities in data. It has applications in statistical data analysis, signal processing, image analysis, information retrieval, bioinformatics, data compression, computer graphics, and

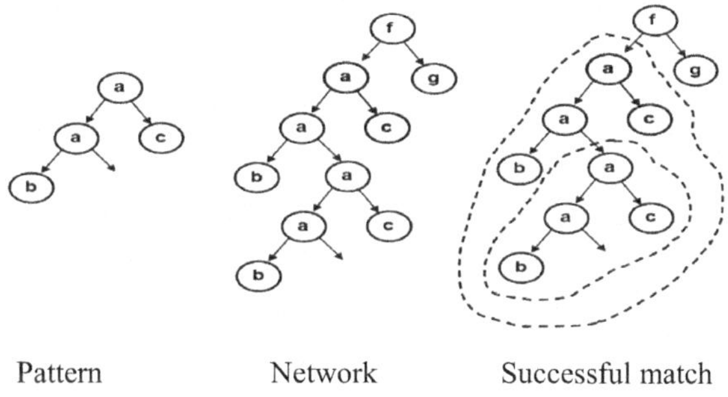

Fig. 9.3 Network pattern matching example

machine learning. Pattern recognition has its origins in statistics and engineering; some modern approaches to pattern recognition include the use of machine learning.

Consciousness as pattern recognition is shown in [89]. This is proof of the strong AI hypothesis that machines can be conscious. It is a phenomenological proof that pattern recognition and subjective consciousness are the same activity expressed using different terms. Therefore, it proves that the essential subjective processes of consciousness are computable and identifies significant features and requirements of a conscious system.

Pattern recognition (psychology) is discussed in [90]. In psychology and cognitive neuroscience, pattern recognition describes a cognitive process that matches information from a stimulus with information retrieved from memory. Pattern recognition occurs when information from the environment is received and entered into short-term memory, causing automatic activation of specific content in long-term memory.

- **Pattern languages**

A pattern language is discussed in [91]. The language description—the vocabulary—is a collection of named, described solutions to problems in a field of interest. These are called design patterns. So, for example, the language for architecture describes items like settlements, buildings, rooms, windows, latches, etc. Each solution includes syntax, a description that shows where the solution fits in a larger, more comprehensive or more abstract design. A pattern language is an organized and coherent set of patterns, each of which describes a problem and the core of a solution that can be used in many ways within a specific field of expertise.

In addition, a *pattern language* is also considered in [92]. The book creates a new language, what the authors call a pattern language derived from timeless entities called patterns. As they write in the introduction, all 253 patterns together form a language. The pattern language is structured as a network, where each pattern may have a statement referenced to another pattern by placing that pattern's number in brackets. This idea of pattern language as a network is symbolically shown in Fig. 9.4.

What is a pattern language is discussed in [93]. In this work, pattern language is an attempt to express the deeper wisdom of what brings aliveness within a particular field of human endeavor, through a set of interconnected expressions arising from

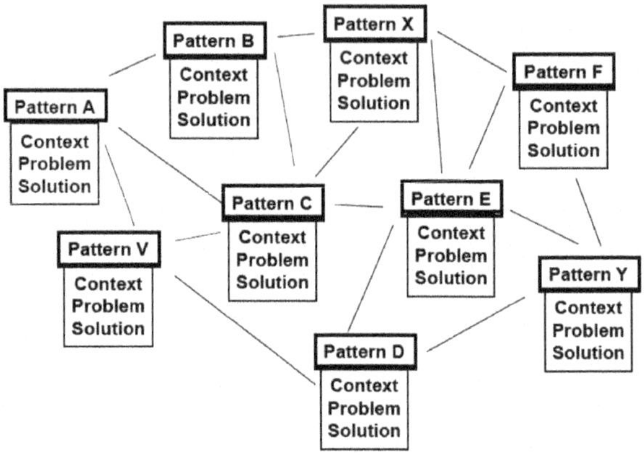

Fig. 9.4 Pattern language as a network

that wisdom. Aliveness is one placeholder term for "the quality that has no name": a sense of wholeness, spirit, or grace, that while of varying form, is precise and empirically verifiable.

"*Pattern language: towns, buildings, and construction*" is studied in [94]. Patterns, the units of this language, are answers to design problems. More than 250 patterns in this pattern language are given: each consists of a problem statement, a discussion of the problem with an illustration, and a solution. Many of the patterns are archetypal, so deeply rooted in the nature of things that it is likely that they will be a part of human nature and human action.

A *pattern language for a pattern language structure* is analyzed in [95]. This paper aims to help the "writers" of pattern languages create better pattern languages. It focuses not on the aesthetics of pattern languages but on their structure: how patterns work together to build a system. The paper assumes that a pattern language is a designed system and, therefore, the theory of system design and evolution underlies the language.

Growing a pattern language (for security) is discussed in [96]. The paper presents a pattern language containing all security patterns that have been published in various places. It describes the mechanism of growing this pattern language: how to catalog the security patterns from books, papers, and pattern catalogs; how to classify the patterns to help developers find appropriate patterns; and how to identify and describe the relationships between patterns in the pattern language.

The idea of *pattern languages as the media for a creative society* is described in [97]. This paper proposes new languages for basic skills in the creative society where people create their own goods, tools, concepts, knowledge, and mechanisms with their own hands: the skills of learning, presentation, and collaboration. These languages are written as a pattern language which is a way of describing the tacit practical knowledge.

Change making *patterns and a pattern language for fostering social entrepreneurship* are presented in [98]. A pattern language named "Change Making Patterns" was created by conducting interviews with social entrepreneurs. The objective of these patterns is to encourage more individuals to take their own actions to make a better world with fewer social problems. The background of the patterns and ideas on how they can be applied for social entrepreneurial education are provided.

Pattern languages in interaction design, structure and organization are discussed in [99]. When individual patterns for interaction design started to appear, the issue of structuring collections of patterns into pattern languages became relevant both from a theoretical and a practical perspective. This paper investigates how pattern languages in interaction design can be structured in a meaningful and practical way. A top-down approach is taken where patterns for interaction design are organized hierarchically.

A *pattern language for pattern writing* is discussed in [100]. As the pattern community has accumulated experience in writing and reviewing patterns and pattern languages, the authors have begun to develop insight into pattern-writing techniques and approaches that are particularly effective at addressing certain recurring problems. This pattern language attempts to capture some of these "best practices" of pattern writing both by describing them in pattern form and by demonstrating them in action.

9.3 Basics of Spatial Grasp Model and Technology

Only basic ideas of this Spatial Grasp paradigm are mentioned here, with more on it in [56–73] and Chaps. 3, 4 and 5. Within Spatial Grasp Model and Technology (SGT), a high-level operational scenario expressed in recursive Spatial Grasp Language (SGL), starting in any world point *propagates, covers, and matches the distributed environment in parallel wavelike mode*, as symbolically shown in Fig. 9.5. Such propagation can result in returning and analyzing the reached states and data, which may be arbitrarily remote, or used for launching more waves.

The SGL allows for *direct space presence and operations* with unlimited powers and parallelism. Its universal recursive organization, with operational scenarios called *grasp*, can be expressed by a single formula:

$$\boldsymbol{grasp} \ \rightarrow \boldsymbol{constant}|\boldsymbol{variable}|\boldsymbol{rule}\ (\{\boldsymbol{grasp},\})$$

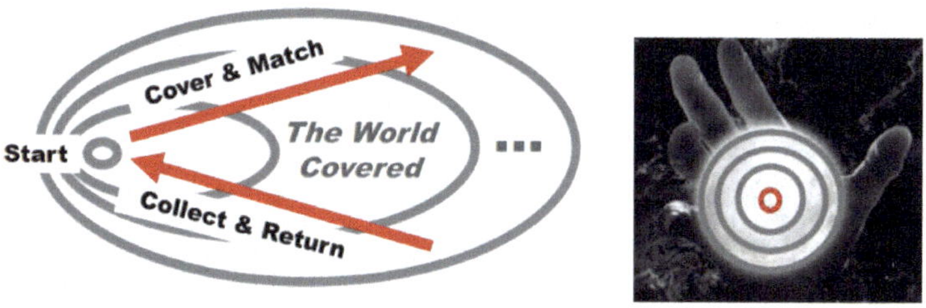

Fig. 9.5 Parallel wavelike world coverage under spatial grasp model

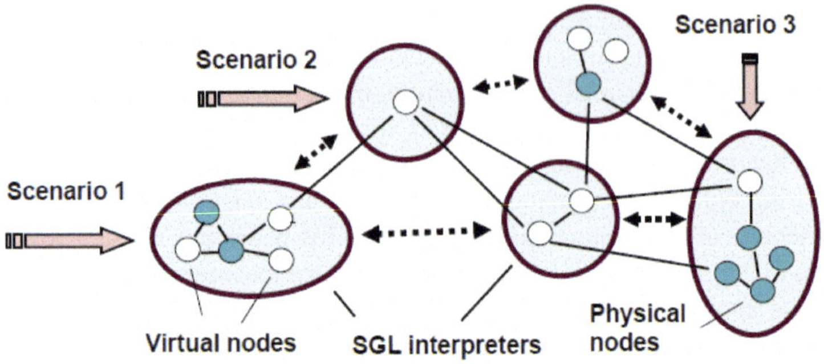

Fig. 9.6 SGL distributed networked interpretation

The *rule* expresses certain action, control, description or context accompanied with operands, which can be any *grasps* too. The rules, starting in certain points, can organize navigation of the world sequentially, in parallel, or any combinations thereof. They can result in the same application points or cause movement to other world points with obtained results left there or returned.

The *SGL interpreter* consists of specialized modules serving SGL scenarios or their parts happened to be inside this interpreter, also organizing exchanges with other interpreters for distributed SGL scenarios. Communicating SGL interpreters can be in arbitrary number of copies, representing *powerful spatial engines operating without central resources or control*, as in Fig. 9.6. They can effectively work with *spatial graph and network data* of any volumes, coverage, and distributions.

9.4 Creating Patterns

- **Regular patterns**

An example of a regular network or graph pattern is shown in Fig. 9.7.

Fig. 9.7 Regular pattern

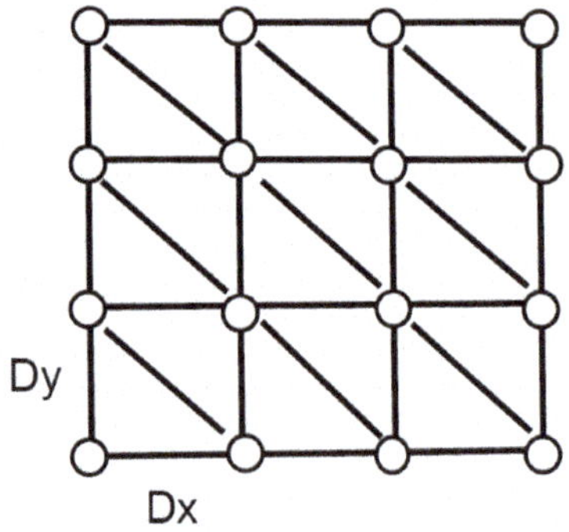

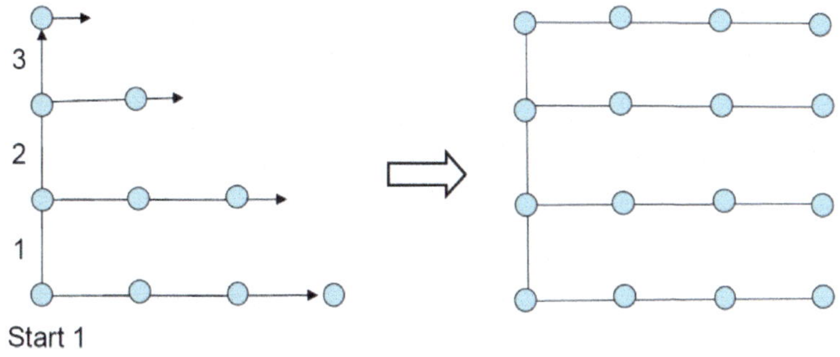

Fig. 9.8 Vertical-horizontal pattern creative growth

(a) *Two step sequential self-growing solution* for its creation in SGL can be as follows.

Step 1: Vertical-horizontal creative growth, see Fig. 9.8.

```
nodal(Start = ...);
frontal(Num = ..., Dx = ..., Dy = ..., Nn = ...,
     Ln = ...);
create_node(Start);
repeat(Num)(
 free_repeat(Num)(create(Ln, (shift(Dx), Nn))));
 create(Ln, (shift(Dy), Nn)))
```

Step 2: Horizontal-vertical linkup growth, see Fig. 9.9.

```
nodal(Start = ...);
frontal(Dx = ..., Dy = ..., Ln = ...);
hop(Start);
repeat(hop(Ln, shift(Dx));
     free_repeat(
      linkup(Ln, shift(Dy)),
      free_linkup(Ln, (shift(Dy); shift(-Dx))))
```

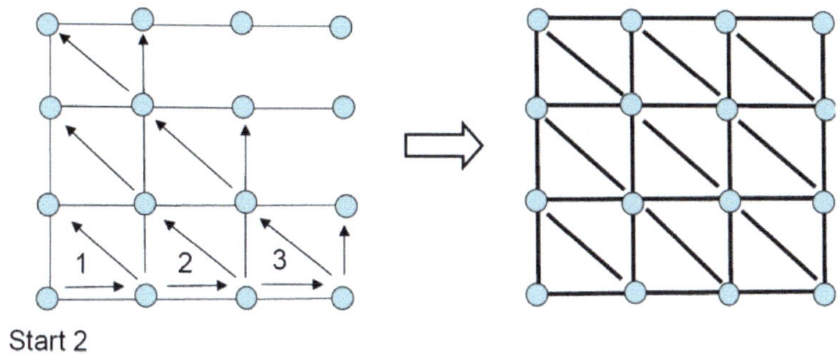

Fig. 9.9 Horizontal-vertical linkup pattern growth

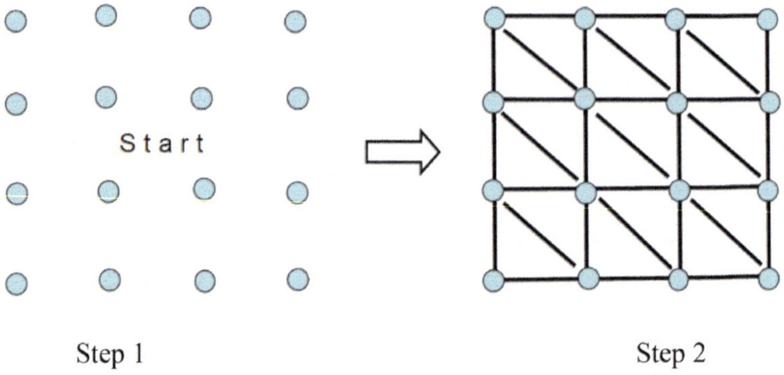

Fig. 9.10 Parallel pattern creation

Full two-step self-growing sequential solution will be as follows:

```
nodal(Start = ...);
frontal(Num = ..., Dx = ..., DY = ..., Nn = ...,
      Ln = ...);
create_node(Start);
stay_repeat(Num)(
  free_repeat(Num)(create(Ln, (shift(Dx), Nn)));
  create(Ln, (shift(Dy), Nn))));
repeat(hop(Ln, shift(Dx));
     free_repeat(
      linkup(Ln, shift(Dy));
      free_linkup(Ln, (shift(Dy); shift(-Dx))))
```

(b) *Full parallel solution of this pattern, which also has two steps*, can be organized as follows, see also Fig. 9.10.

```
nodal(Xmin = ..., Ymin = ...; Dx = ...,
    Dy = ..., Number = ..., Ycurr, Xcoord);
frontal(Xcurr, Ycoord, Ln = ...);
sequence(
  (Xcoord = array(Xmin, Dx, Number);
  Ycoord = array(Ymin, Dy, Number);
  parallel(split(Xcoord); Xcurr = VALUE);
  parallel(split(Ycoord); Ycurr = VALUE);
  create_node(Xcurr, Ycurr)),
  (hop_nodes(all);
  parallel(
   linkup(Ln, shift(Dx)),
   linkup(Ln, shift(Dy)),
   linkup(Ln, (shift(Dx); shift(-Dy)))))
```

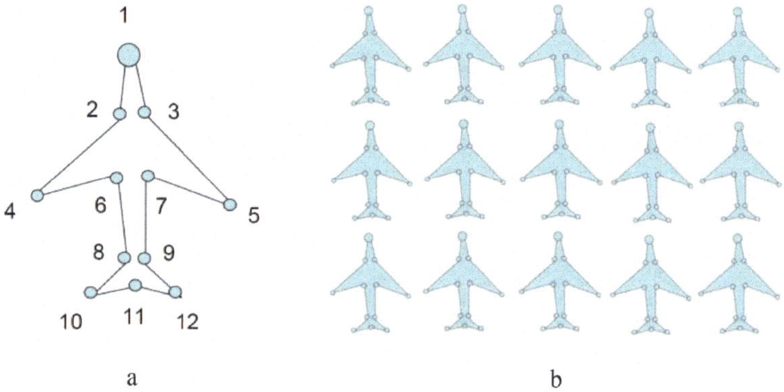

Fig. 9.11 Image pattern (**a**) and its regular replication (**b**)

- **Patterns of concrete objects**

Pattern as a symbolic aircraft image is shown in Fig. 9.11a and its regular replication in Fig. 9.11b.

Implementation of this pattern in SGL will be as follows.

(a) *Single image pattern*, as in Fig. 9.11a.

```
frontal(N = ((dx1_dy1), (dx2_dy2), ..., (dx12_dy12)),
     Ln = ...);
nodal(Start) = X1_Y1;
move(Start); N = (split(N); VALUE + WHERE);
create(N:1; (Ln, N:2); (Ln, N:4); (Ln, N:6);
     (Ln, N:8); (Ln, N:10); (Ln, N:11);
     (Ln, N:12); (Ln, N:9); (Ln, N:7);
     (Ln, N:5); (Ln, N:3));
linkup(Ln, N:1)
```

(b) *Multiple image pattern*, as in Fig. 9.11b.

```
frontal(N = ((dx1_dy1), (dx2_dy2), ..., (dx12_dy12)),
     Ln = ...);
nodal(Start) = (X1_Y1, X2_Y2, X3_Y3, ..., XM_YM);
move_parallel(Start); N = (split(N); VALUE + WHERE);
create(N:1; (Ln, N:2); (Ln, N:4); (Ln, N:6);
     (Ln, N:8); (Ln, N:10); (Ln, N:11);
     (Ln, N:12); (Ln, N:9); (Ln, N:7);
     (Ln, N:5); (Ln, N:3));
linkup(Ln, N:1)
```

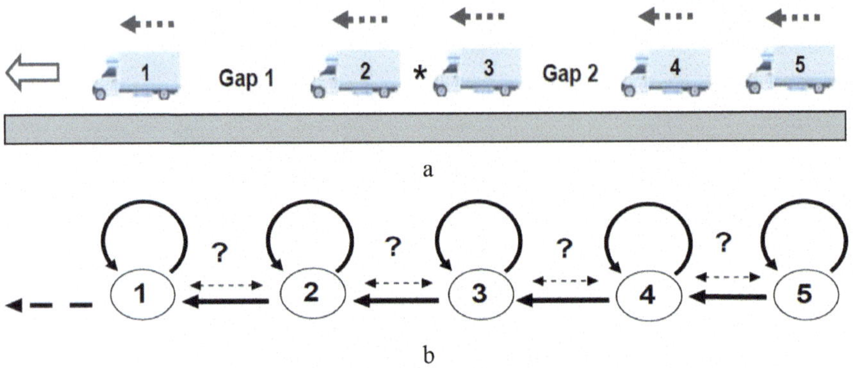

Fig. 9.12 Pattern of moving column (**a**) with correcting distances between vehicles (**b**)

9.5 Pattern-Based Practical Solutions

9.5.1 *Managing Transport Column (Dynamic Chain Pattern)*

Starting in all vehicles of the column, the SGL scenario, if needed, tries to reduce or increase distances between neighboring vehicles, using existing threshold gap standards, as in Fig. 9.12a, b.

```
frontal(Ln = ..., Thresh1 = ..., Thresh2 = ...,
      Diff = ..., Speed = ..., Pos);
sequence(
 create(node(1, W1); (Ln, node(2, W2));
      (Ln, node(3, W3)); (Ln, node(4, W4));
      (Ln, node(5, W5))),
 (hop_nodes(all); SPEED = Speed;
  repeat_stay(
   Pos = WHERE; hop_link(-Ln);
   Dist = distance(Pos, WHERE);
   if(Dist < Thresh1, reduce(SPEED, Diff));
   if(Dist > Thresh2, increase(SPEED, Diff)))))
```

9.5.2 *Collecting Zone Coordinates (Area Surrounding Pattern)*

The following scenario, starting at some point near the distributed area, collects its border coordinates by moving around it in close distance each time, finally outputting the full border coordinates, See Fig. 9.13.

```
frontal(Start = ..., Step = ..., Near = ..., Coord);
go(Start);
```

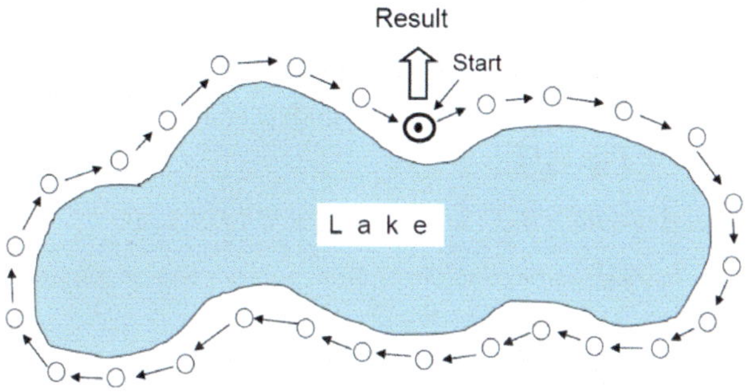

Fig. 9.13 Pattern as zone coordinates

```
repeat(go(right, Step, Near); append(Coord, WHERE);
    if(distance(WHERE, Start) <= Step,
      done_output(Coord)))
```

9.5.3 *Following Mobile Objects (Collective Tracking Pattern)*

The following self-evolving spatial scenario tracks and controls complex objects propagating via a distributed network of radar stations (see Fig. 9.14).

```
hop(all_nodes); frontal(Object);
whirl(
 Object = search(aerial, new));
 visibility(Object) > threshold;
 repeat(
  loop(visibility(Object) > Threshold);
  max_destination(hop(all_neighbors);
           visibility(Object))))
```

Fig. 9.14 Pattern of tracking moving objects

9.6 Pattern Recognition

The following scenarios recognize proper nodes and peculiar structures in the distributed network of Fig. 9.15.

(a) *Finding all nodes with 5 links to other nodes* (Fig. 9.16a):

```
output(hop_nodes(all);
    count(hop_links(all)) == 5; NAME)
```

Answer: F, D, B. I.

(b) *Finding all cliques with 4 nodes* (Fig. 9.16b):

```
hop_nodes(all);
frontal(Clique = NAME, Number = 4);
repeat(
  hop_links(all); notbelong(NAME, Clique);
  true(and_parallel(
      hop(link(any), nodes_all(Clique))));
  append(Clique, NAME));
if(count(Clique) == Number, output(Clique))
```

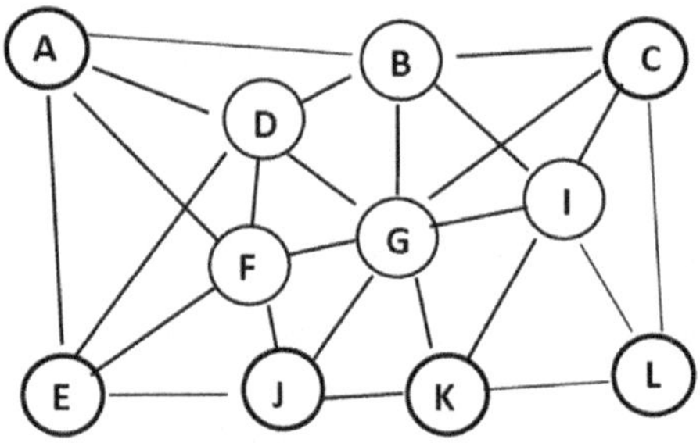

Fig. 9.15 A distributed network example

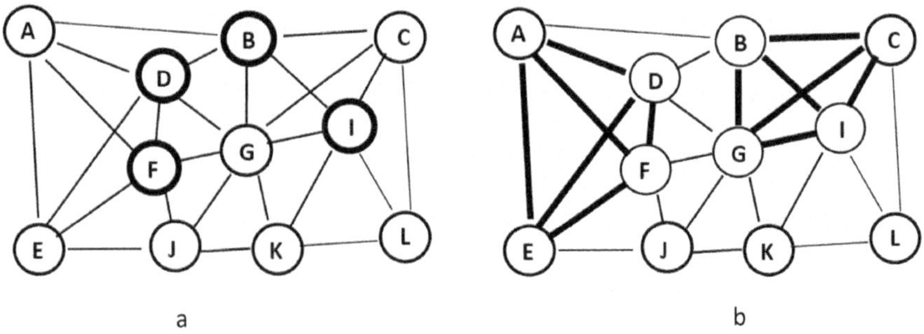

Fig. 9.16 Finding proper nodes (**a**) and proper structure as clique (**b**)

9.7 Pattern Matching

9.7.1 Arbitrary Graph Pattern

The following SGL scenario finds an exact match of the given pattern of Fig. 9.17a with a distributed network of Fig. 9.17b, with matching template (with fat arrows) as a path through all pattern nodes.

```
hop_nodes(all); Match = NAME;
repeat(6)(hop_links(all); notbelong(NAME, Match);
      append(Match, NAME);
      true(hop(links(any), node(Match[1]))));
hop(links(any), node(Match[2])); output(Match)
```

Reply (one of): W-X-Z-V-U-T-Y: G, B, I, K, J, F, D

Another pattern, with the template as a path through all its nodes too, is shown in Fig. 9.18.

```
hop_nodes(all); Match = NAME;
repeat(3)(hop_links(all); notbelong(NAME, Match);
```

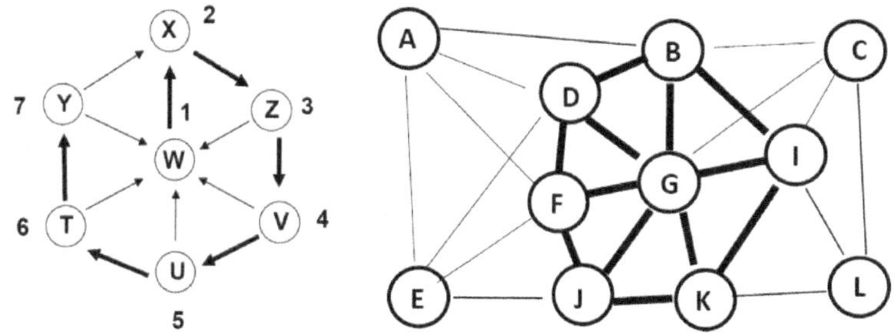

a b

Fig. 9.17 Graph pattern with its matching template (**a**), and successful network match (**b**)

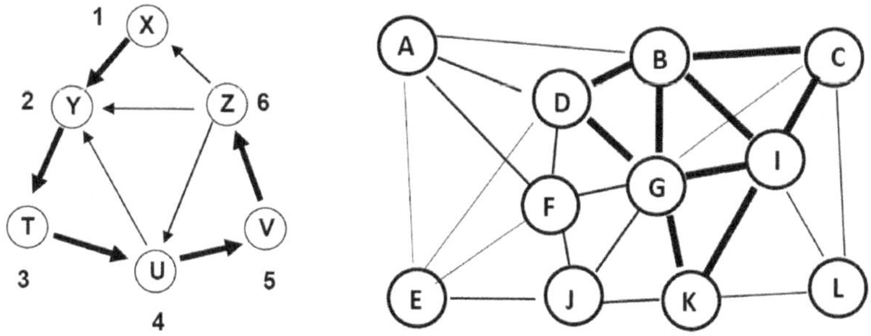

Fig. 9.18 Another pattern-template (**a**) and one of its network matches (**b**)

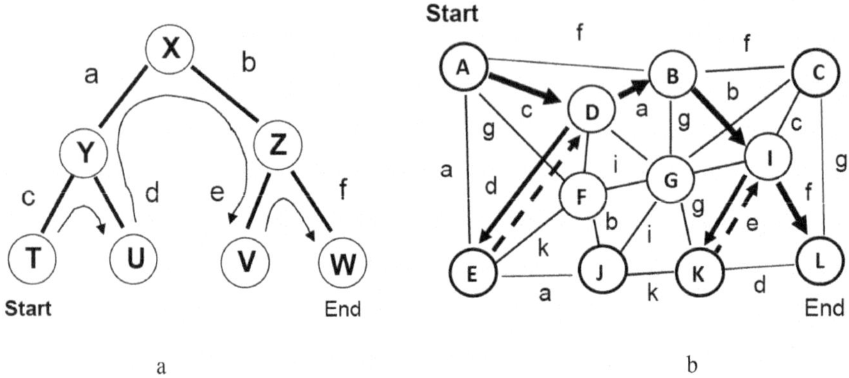

Fig. 9.19 Using path through all pattern nodes (**a**), matching result (**b**)

```
          append(Match, NAME);
    true(hop(links(any), node(Match[2]));
    repeat(2)(hop_links(all); notbelong(NAME, Match);
          append(Match, NAME);
    true(hop(links(any), and(nodes(Match[1,2,4]))));
    output(Match)
  output(Match)
```

Reply (one of): X-Y-T-U-V-Z: B, D, G, K, I, C

9.7.2 Tree Pattern

(a) *With template as a path through all pattern nodes*, as in Fig. 9.19.

```
hop_nodes(all); frontal(Match) = NAME;
hop_link(c); append(Match, NAME);
hop_link(d); append(Match, NAME); hop(BACK);
hop_link(a); append(Match, NAME);
hop_link(b); append(Match, NAME);
hop_link(e); append(Match, NAME); hop(BACK);
hop_link(f); append(Match, NAME;
output(Match)
```

Reply: T-Y-U-X-Z-V-W: A, D, E, B, I, K, L

(b) *With top-down template*, see Fig. 9.20.

Unordered, parallel solution:

```
hop_nodes(all);
output(and(
  (hop_link(a);
   and(hop_link(c), hop_link(d), NAME)),
```

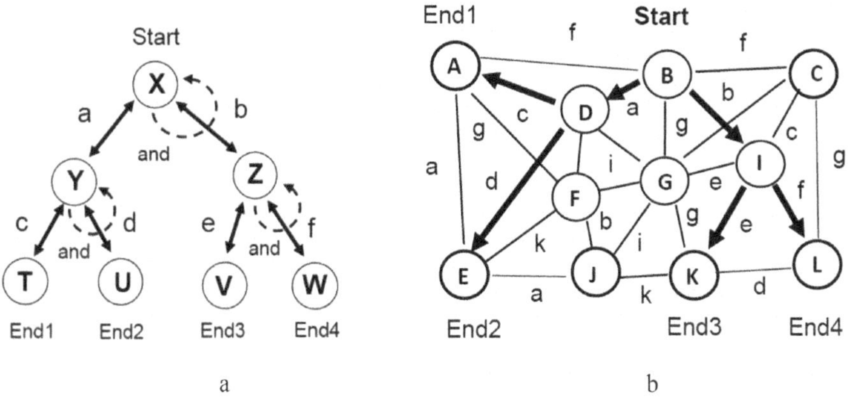

Fig. 9.20 Tree-like template (a) and its top-down parallel match (b)

```
(hop_link(b);
 and(hop_link(e), hop_link(f), NAME)),
NAME))
```

Possible reply: B, I, L, K, D, E, A

Ordered, sequential solution:

```
hop_nodes(all);
output(and_seq(
 (hop_link(a);
  and_seq(hop_link(c), hop_link(d), NAME)),
 (hop_link(b);
  and_seq(hop_link(e), hop_link(f), NAME)),
NAME))
```

Reply: (T-U-Y)-(V-W-Z)-X: (A, E, D), (K, L, I), B

9.8 SGL as a Real Pattern Language

After the detailed analyses, practical implementation and use of pattern-based solutions in many areas [74–100], SGL cannot not be treated just as a networked set of specific patterns like in the existing pattern languages [91–100]. On the contrary, it reflects the universal paradigm *capable of expressing and using any individual or general patterns and their systems*, whether passive or active, as well as combining descriptive patterns with any centralized and distributed processes. A summary of this conclusion may be as follows, which can confirm that SGL can represent a real and effective pattern language with practical applications in numerous areas.

Descriptive patterns. Any pattern parts, their compositions, and structures, with local or global parameters, can be naturally described in SGL and subsequently used for pattern recognition and matching by special processes.

Creative patterns. Patterns describing the composition and structures of different images, whether regular or arbitrary, can themselves be active parallel processes capable of creating or drawing such images autonomously in distributed spaces.

Patterns as spatial processes. These can provide any investigation, control, management, and simulation of any system, which may have a terrestrial or celestial nature. They can also express and simulate large distributed spatial processes like the spread of global viruses and fighting them.

Pattern recognition. This can be effectively organized in any physical, virtual, or combined system where the pattern recognition process relies on detailed or generalized pattern qualification and representation data expressed in logical, numerical, or structural form.

Self-matching patterns. Any graph-based pattern can be organized as active self-matching templates autonomously evolving and spreading in distributed spaces in a wave-like parallel manner. They can return names, addresses, and/or physical coordinates of the found elements and their structures, as well as influence the matched systems.

Combined patterns. Any combinations of the above-mentioned and following patterns, of any complexity, that can recognize, match, analyze, change, and control any distributed systems can be effectively organized. These patterns may virtually or physically migrate and fly over the covered systems in a controlled virus-like mode.

Cooperating and conflicting patterns. Different self-powered patterns may compete or cooperate in distributed systems and spaces, for example, representing collectives working on common problems or reflecting opposing units on a battlefield competing and destroying each other.

Psychological patterns. As it was already shown and discussed in SGT-related publications, like [60, 62, 65, 68, 69, 71] and others, the main gestalt psychology ideas, laws, and examples, based on mentally grasping the whole of concepts and events first, can be effectively represented and modeled with the use of spatial patterns in SGL.

Recursive patterns. Patterns of any complexity can be effectively organized in SGL using its universal recursive structure, with any above-mentioned pattern capable of including any other patterns, each of which can do the same, and to an unlimited depth.

9.9 Conclusions

This concludes with a strong belief that SGT reflects the universal paradigm capable of expressing, creating, and using any individual or general patterns and their systems, and SGL can be used as a real, effective, and very compact language for pattern representation and operations, truly named as "pattern language". SGT can believably contribute to pattern theory and resultant technologies as a universal model treating the whole world, whether physical, virtual, psychological, or mental, as spatial patterns. The solutions provided and discussed in this chapter were obtained in a spatial thinking, spatial pattern recognition, and spatial pattern matching mode, rather than in traditional logic-based and algorithmic philosophy and culture. Moreover, being highly parallel and fully distributed, they challenge existing opinions that parallel and distributed algorithms are usually much more complex than traditional sequential ones for solving the same problems, just proving the opposite, and especially for the pattern and network-related problems.

References

1. Transport networks. Inspire Registry, (2015). https://inspire.ec.europa.eu/theme/tn
2. Transportation networks. Travel Forecasting Resource. https://tfresource.org/topics/Transportation_networks.html
3. Durkin, J., Goodman, J., et al.: Building multiservice transport networks (Networking Technology), 1st edn. Cisco Press, (2012)
4. Murakami, M., Li, H., Ryoo, J-D.: Packet transport networks: overview and future direction. In: 2 nd APT/ITU conformance and interoperability workshop (C&I–2), (2014). https://www.itu.int/en/ITU-T/C-I/interop/Documents/20140826/CI-2_INP-09_Packet_Transport_Networks_Overview_and_Future_Direction.pdf
5. Air Transport Network. https://en.wikipedia.org/wiki/Air_Transport_Network
6. Qureshi, K.N., Ul Islam, F., et al.: Improved road segment-based geographical routing protocol for vehicular ad-hoc networks. Electronics **9**(8), 1248 (2020). https://doi.org/10.3390/electronics9081248. https://www.mdpi.com/2079-9292/9/8/1248
7. Sujan.: What is communication network? definition, importance, and 4 types (Explained), (2023). https://tyonote.com/communication_network/
8. Fitzek, F.H.P., Granelli, F., et al.: Computing in communication networks: from theory to practice, 1st edn .Academic Press, (2020)
9. Srikant, R., Ying, L.: Communication networks: an optimization, control, and stochastic networks perspective, 1st edn. Cambridge University Press, (2013)
10. Raghunandan, K.: Introduction to wireless communications and networks: a practical perspective (textbooks in telecommunication engineering), 1st ed., 2022 edition. Springer, (2023)
11. Lou, Y., Ahmed, N.: Underwater communications and networks (textbooks in telecommunication engineering), Springer, (2021)
12. Kaufman, C., Perlman, R., et al.: Network security: private communication in a public world (prentice hall series in computer networking and distributed systems), 3rd edn. Addison-Wesley Professional, (2022)
13. Wu, Y., Singh, S., et al.: 6G mobile wireless networks (computer communications and networks), 1st edn. 2021 edition. Springer, (2022)
14. Social network. https://en.wikipedia.org/wiki/Social_network
15. Kenton, W.: What is social networking? Investopedia. Updated May 24, 2023. https://www.investopedia.com/terms/s/social-networking.asp#:~:text=Social%20networking%20refers%20to%20using,)%2C%20Instagram%2C%20and%20Pinterest
16. Nanawati, S.: Social network analytics, published in analytics vidhya. (2019). https://medium.com/analytics-vidhya/social-network-analytics-f082f4e21b16
17. Singh, C.: 20+ most popular social networking sites in 2024. (2023). https://www.socialpilot.co/social-networking-sites
18. Jackson, M.O.: Social and economic networks, Illustrated edition. Princeton University Press, (2010)
19. Harrison, T.: Battlenetworks and the future force, center for strategic and international studies. (2021). https://www.csis.org/analysis/battle-networks-and-future-force

20. Stillion, J., Clark, B.: What it takes to win: succeeding in 21st century battlenetwork competitions. Center for Strategic and Budgetary Assessments (CSBA), (2015). https://csbaonline.org/uploads/documents/What-it-Takes-to-Win.pdf
21. Hadley, G.: DAF outlines a new 'battle network' as its contribution to JADC2. Air & Space Forces Mag, (2023). https://www.airandspaceforces.com/daf-battle-network-contribution-jadc2/
22. The challenges of military networks and technological developments, Elbit Systems. https://elbitsystems.com/e-lynx_multi_channel/
23. Themnér, A., Karlén, N.: Building a safety net: explaining the strength of ex-military networks. Secur. Stud., **29**(2), 268–300 (2020). https://doi.org/10.1080/09636412.2020.1722851 https://www.tandfonline.com/doi/full/10.1080/09636412.2020.1722851
24. What is an economic network? Advantages and disadvantages of economic networks, Vietnambiz, (2019). https://vietnambiz.vn/mang-luoi-kinh-te-economic-network-la-gi-uu-nhuoc-diem-cua-mang-luoi-kinh-te-20191011113428212.htm
25. Halton, C.: Economic network: what it is, pros and cons, examples. Updated October 24, 2022. https://www.investopedia.com/terms/e/economic-network.asp#:~:text=Key%20Takeaways-,An%20economic%20network%20is%20a%20combination%20of%20individuals%2C%20groups%2C%20or,companies%20or%20partnerships%20between%20corporations
26. Todo, Y., Matous, P.: Economic networks for more innovative and resilient economies. (2015). https://cepr.org/voxeu/columns/economic-networks-more-innovative-and-resilient-economies
27. Emmert-Streib, F., Tripathi, S., et al.: Understanding the world economy in terms of networks: a survey of data-based network science approaches on economic networks. Front. Appl. Math. Stat., Sec. Mathematical Finance, **4**, (2018). https://doi.org/10.3389/fams.2018.00037. https://www.frontiersin.org/articles/10.3389/fams.2018.00037/full
28. Wixted, B.: Innovation system frontiers: cluster networks and global value (advances in spatial science). Springer, (2009)
29. Manyika, J.: What the economy looks like so far for individuals in the 21st century. (2020). https://www.linkedin.com/pulse/what-economy-looks-like-so-far-individuals-21st-century-james-manyika
30. Hayes, A.: Distribution network: definition, how it works, and examples, investopedia. Updated February 22, 2021. https://www.investopedia.com/terms/d/distribution-network.asp#:~:text=In%20a%20supply%20chain%2C%20a,or%20through%20a%20retail%20network
31. Distribution Network. The flow of goods from a producer or supplier to an end consumer. Written by CFI Team. https://corporatefinanceinstitute.com/resources/valuation/distribution-network/
32. What are types of distribution networks in supply chain management? oboloo FAQ's. https://oboloo.com/blog/what-are-types-of-distribution-networks-in-supply-chain-management/
33. Distribution Network—What is it and what are the benefits? https://www.salesbabu.com/blog/what-is-a-distribution-network-and-its-benefits/
34. Wu, Q., Shen, F., et al.: Optimal operation of active distribution networks: congestion management. Academic Press, Voltage Control and Service Restoration (2023)
35. Watts, S.: What is a virtual network? BMC, (2020). https://www.bmc.com/blogs/virtual-network/
36. What is Virtual Networking? VMware by Broadcom. https://www.vmware.com/topics/glossary/content/virtual-networking.html
37. Network virtualization. https://en.wikipedia.org/wiki/Network_virtualization
38. Carlos, O., Duarte, M.B., Pujolle, G.: Virtual networks: pluralistic approach for the next generation of internet (networks and telecommunications), 1st edn. Wiley-ISTE, (2013)
39. Fundamentals of Virtual Networking. https://www.geeksforgeeks.org/fundamentals-of-virtual-networking/
40. Neural network. https://en.wikipedia.org/wiki/Neural_network

41. Anand, V.: Introduction to neural network in machine learning, Updated On August 10th, 2023. https://www.analyticsvidhya.com/blog/2022/01/introduction-to-neural-networks/
42. Ashtari, H.: What is a neural network? Definition, working, types, and applications in 2022. Last Updated: August 3, 2022. https://www.spiceworks.com/tech/artificial-intelligence/articles/what-is-a-neural-network/
43. Artificial Neural Network. NVIDIA Developer. https://developer.nvidia.com/discover/artificial-neural-network
44. Wu, L., Cui, P., et al.: Graph neural networks: foundations, frontiers, and applications, 1st ed, 2022 edition. Springer, (2022)
45. Borsboom, D., Deserno, M.K., Rhemtulla, M. et al.: Network analysis of multivariate data in psychological science. Nat Rev Methods Primers. **1**, 58 (2021). https://doi.org/10.1038/s43586-021-00055-w. https://www.nature.com/articles/s43586-021-00055-w
46. Jones, P.J., Mair, P., McNally, R.J.: Visualizing psychological networks: a tutorial in R. Front. Psychol, Sec. Quant. Psychol. Meas., **9**, (2018). https://doi.org/10.3389/fpsyg.2018.01742. https://www.frontiersin.org/articles/10.3389/fpsyg.2018.01742/full
47. Fonseca-Pedrero, E.: Network analysis in psychology. Psychol. Pap., **39**(1), 1–12 (2018). https://doi.org/10.23923/pap.psicol2018.2852. https://www.papelesdelpsicologo.es/pdf/2852.pdf
48. Bringmann, L.F.: Person-specific networks in psychopathology: Past, present, and future. Curr. Opin. Psychol., **41**, 59–64 (2021). https://www.sciencedirect.com/science/article/pii/S2352250X21000300?via%3Dihub
49. Yuqing, C., Shuyang, D. et al.: Network analysis and its applications in psychology. Adv. Psychol. Sci., **28**(1), 178–190 (2020). https://doi.org/10.3724/SP.J.1042.2020.00178. https://journal.psych.ac.cn/adps/EN/10.3724/SP.J.1042.2020.00178
50. Campana, P.: Explaining criminal networks: Strategies and potential pitfalls. Methodol. Innov. **9**, 1–10 (2016). https://doi.org/10.1177/2059799115622748. https://journals.sagepub.com/doi/pdf/10.1177/2059799115622748
51. Bichler, G.: Understanding criminal networks: a research guide, 1st edn. University of California Press, (2019)
52. Morselli, C.: Crime and networks (Criminology and Justice studies). Routledge, (2013)
53. New forms of organized crime: networked structure. UNODS, Unite Nations Office on Drugs and Crime. https://www.unodc.org/e4j/en/organized-crime/module-7/key-issues/networked-structure.html
54. Hufnagel, S., Moiseienko, A.: Criminal networks and law enforcement: global perspectives on illegal enterprise (Transnational Criminal Justice), 1st edn. Routledge, (2019)
55. Detecting and disrupting transnational criminal organizations: analytics for interdependent smuggling and money-laundering networks, criminal investigation and network analysis. DHS Centers of Excellence. https://cina.gmu.edu/projects/detecting-and-disrupting-transnational-criminal-organizations-analytics-for-interdependent-smuggling-and-money-laundering-networks
56. Sapaty, P.S.: A distributed processing system. European Patent N 0389655, Publ. 10.11.93, European Patent Office
57. Sapaty, P.S.: Mobile processing in distributed and open environments. John Wiley & Sons, New York (1999)
58. Sapaty, P.S.: Ruling distributed dynamic worlds. John Wiley & Sons, New York (2005)
59. Sapaty, P.S.: Managing distributed dynamic systems with spatial grasp technology. Springer, (2017)
60. Sapaty, P.S.: Holistic analysis and management of distributed social systems. Springer, (2018)
61. Sapaty, P.S.: Complexity in international security: a holistic spatial approach. Emerald Publishing, (2019)
62. Sapaty, P.S.: Symbiosis of real and simulated worlds under spatial grasp technology. Springer, (2021)
63. Sapaty, P.S.: Spatial grasp as a model for space-based control and management systems. CRC Press, (2022)

64. Sapaty, P.S.: The spatial grasp model: applications and investigations of distributed dynamic worlds. Emerald Publishing, (2023)
65. Sapaty, P.S.: Providing integrity, awareness, and consciousness in distributed dynamic systems. CRC Press, (2025)
66. Sapaty P.S.: Spatial management of air and missile defence operations. Math. Mach. Syst. (1), (2023). http://www.immsp.kiev.ua/publications/articles/2023/2023_1/01_23_Sapaty.pdf
67. Sapaty, P.S.: Providing distributed system integrity under spatial grasp technology. Math. Mach. Syst. (2), (2023). http://www.immsp.kiev.ua/publications/articles/2023/2023_2/02_23_Sapaty.pdf
68. Sapaty, P.S.: Providing global awareness in distributed dynamic systems. Int. Relat. Dipl., **11**(2), 87–100 (2023). https://doi.org/10.17265/2328-2134/2023.02.002. https://www.davidpublisher.com/Public/uploads/Contribute/6486c3d05a6cc.pdf
69. Sapaty, P.S.: Simulating distributed consciousness with spatial grasp model. Math. Mach. Syst. (3), (2023)
70. Sapaty, P.S.: Network centrality operations under spatial grasp technology. J. Adv. Artif. Intell. Mach. Learn., **1**(1), 1–11 (2023). https://www.scivisionpub.com/journals/articleinpress-journal-of-advances-in-artificial-intelligence-and-machine-learning
71. Sapaty, P.S.: Managing distributed systems with spatial grasp patterns. Math. Mach. Syst. (4), (2023)
72. Sapaty, P.S.: Spatial networks as models for organoid cultures and brain research. Math. Mach. Syst. (1), (2024)
73. Sapaty, P.S.: Basic network operations under spatial grasp model. Sci Set J Physics **3**(1), 01–16 (2024)
74. Pattern. https://en.wikipedia.org/wiki/Pattern
75. Grenander, U., Miller, M.: Pattern theory: From representation to inference, p. 596. Oxford University Press, (2007)
76. de Souza, F.D.M., Sarkar, S., Srivastava, A., Su, J.: Pattern theory-based interpretation of activities. In: 22nd International conf on pattern recognition. (2014). https://projet.liris.cnrs.fr/imagine/pub/proceedings/ICPR-2014/data/5209a106.pdf
77. Mumford, D.: Pattern theory: A unifying perspective. https://www.dam.brown.edu/people/mumford/vision/papers/1994c-96--PattThUnifyingPersp-NC.pdf.
78. Pattern theory. https://en.wikipedia.org/wiki/Pattern_theory
79. Joshi, U.: Patterns of distributed systems. https://martinfowler.com/articles/patterns-of-distributed-systems/
80. Sankey, T.: Statistical descriptions of spatial patterns. Encyclopedia of GIS. Springer, Cham, (2017). https://link.springer.com/referenceworkentry/10.1007/978-3-319-17885-1_1351. https://doi.org/10.1007/978-3-319-17885-1_1351
81. Keim, D.A., Kröger, P., Ankerst, M.: Recursive pattern: A technique for visualizing very large amounts of data. In: IEEE Xplore: Conf visualization. (1995). https://www.researchgate.net/publication/3618236_Recursive_pattern_A_technique_for_visualizing_very_large_amounts_of_data. https://doi.org/10.1109/VISUAL.1995.485140
82. Goertzel, B.: The hidden pattern: A patternist philosophy of mind, p. 424. Brown Walker Press, (2006)
83. Russell, N., ter Hofstede, A.H.M., van der Aalst, W.M.P., Mulyar, N.: Workflow control-flow patterns: a revised view, p. 134. BPMcenter, (2006)
84. van der Aalst, W.M.P., der Hofstede, A.H.M., Kiepuszewski, B., Barros, A.P.: Workflow patterns, (2002). http://www.workflowpatterns.com/documentation/documents/wfs-pat-2002.pdf
85. Pattern matching. https://en.wikipedia.org/wiki/Pattern_matching
86. Cheng, J., Xu Yu, J., Ding, B., Yu, P.S., Wang, H.: Fast graph pattern matching. https://www.microsoft.com/en-us/research/wp-content/uploads/2016/02/icde08gsearch.pdf
87. Reza, T.: Pattern matching in massive metadata graphs at scale. University of British Columbia, (2019). https://open.library.ubc.ca/soa/cIRcle/collections/ubctheses/24/items/1.0387453. https://doi.org/10.14288/1.0387453

88. Pattern recognition. https://en.wikipedia.org/wiki/Pattern_recognition
89. De Walker, V.R.: Consciousness is pattern recognition, computer science>Artificial Intelligence, Copyright 2016. https://arxiv.org/abs/1605.03009
90. Pattern recognition (psychology). https://en.wikipedia.org/wiki/Pattern_recognition_(psychology)
91. Pattern language. https://en.wikipedia.org/wiki/Pattern_language
92. A pattern language: towns, buildings, construction, 1977 book. https://en.wikipedia.org/wiki/A_Pattern_Language
93. What is a Pattern Language? https://groupworksdeck.org/pattern-language
94. Alexander, C.H., Ishikawa, S., Silverstein, M., Jacobson, M., Fiksdahl-King, I., Angel, S.H.: A pattern language: towns, buildings, construction, p. 1171. Oxford University Press, (1977). https://www.amazon.com/Pattern-Language-Buildings-Construction-Environmental/dp/0195019199
95. Winn, T., Calder, P.: A pattern language for pattern language structure. In: Third Asian Pacific conference on pattern languages of programs. KoalaPLoP, (2002). https://www.researchgate.net/publication/228944468_A_pattern_language_for_pattern_language_structure
96. Hafiz, M., Adamczyk, P., Johnson, R.: Growing a pattern language (for Security). In: Procedings of the ACM international symposium on new ideas, new paradigms, and reflections on programming and software. (2012). https://www.researchgate.net/publication/262211609_Growing_a_pattern_language_for_security. https://doi.org/10.1145/2384592.2384607
97. Iba, T.: Pattern languages as media for the creative society. J. Inf. Process. Management. **55** (12), (2013). https://www.researchgate.net/publication/255484746_Pattern_Languages_as_Media_for_the_Creative_Society. https://doi.org/10.1241/johokanri.55.865
98. Shimomukai, E., Nakamura, S., Iba, T.: Change making patterns. a pattern language for fostering social entrepreneurship, p. 112. CreativeShift Lab, (2015)
99. van Welie, M., van der Veer, G.: Pattern languages in interaction design: structure and organization. (2011). https://www.researchgate.net/publication/228881522_Pattern_languages_in_interaction_design_Structure_and_organization
100. Meszaros, G., Doble, J.: A pattern language for pattern writing. The Hillside Group. https://hillside.net/index.php/a-pattern-language-for-pattern-writing

Chapter 10
Networks as Models of Organoid Cultures and Brain Research

10.1 Introduction

The chapter is devoted to the phenomenon called *organoid*, which is of enormously growing interest, importance, and popularity for the last years. An organoid is a miniaturized and simplified version of an organ produced in vitro that mimics the key functional, structural and biological complexity of that organ. Constructed from both normal and diseased tissues, organoids serve as a valuable tool for fundamental research, disease modeling, drug screening and discovery, also in regenerative medicine. For example, brain organoids address fundamental questions in human brain research, whereas tumor organoids are cultured to accelerate finding more precise therapies to fight cancer.

The aim of this chapter is to investigate potential applicability of the developed Spatial Grasp Model and Technology (SGT) based on active distributed self-evolving networking, also its experience of dealing with different types of networks [1–55] discussed in previous chapters, to simulate and enrich the existing and planned works on organoids. Such symbiosis could effectively combine the laboratory tissue growth with intelligent control and management capabilities guaranteed by SGT, contributing altogether into this extremely important area. The rest of the paper is organized as follows.

Section 10.2 reviews existing organoid publications, including of what is organoid, latest organoid books, examples of cells-based techniques investigated in this paper, like cell differentiation and tissue, shapes of organoids, worms, and killer cells. Section 10.3 briefs the developed Spatial Grasp Model and Technology (SGT) described in detail in Chaps. 3, 4 and 5. Section 10.4 provides examples of networking solutions under SGT, like creation, movement, reshaping of nodes, creation of networks in virtual and physical spaces, finding paths in the network, and finding proper nodes. Section 10.5 shows expression of organoid models under SGT, which include cells growth, division and replication, worm creation and moving, killer cell operation, full organoid model and its operation, also explains how investigate

P. S. Sapaty, *Spatial Networking in the United Physical, Virtual, and Mental World*, Studies in Systems, Decision and Control 533,
https://doi.org/10.1007/978-3-031-62154-3_10

parameters of the growing organoids. Section 10.6 describes some other organoid developments which are of interest for their simulation under SGT, which include mini-brains in robotics, organoids emitting brain waves, and creation of organoids in outer space. Section 10.7 concludes the chapter confirming applicability of the developed networking paradigm to express and simulate individual and multiple cells growth up to the whole organoids. References include reviewed networking and organoid publications, also those related to SGT.

10.2 Organoid Publications

We will be providing here a review of organoid definition, classification, research, reproduction and simulation in some latest organoid-related publications. Also reviewing a number of existing cells-based growth, differentiation, communication, and shaping techniques to be analyzed and modeled later in this paper.

10.2.1 What is Organoid

An *organoid* [56] is a miniaturized and simplified version of an organ produced in vitro in three dimensions that mimics the key functional, structural and biological complexity of that organ. They are derived from one or a few cells from a tissue, embryonic stem cells or induced pluripotent stem cells, which can self-organize in three-dimensional culture owing to their self-renewal and differentiation capacities.

Organoids are anatomically and functionally close to the organs in the living body [57]. They can be constructed from both normal and diseased tissues, allowing for utilization in various research applications. These tissue structures are used for basic research such as embryology, physiology, and evolution as well as more involved research including characterizing pathological disease conditions and drug discovery research. See Fig. 10.1 for examples.

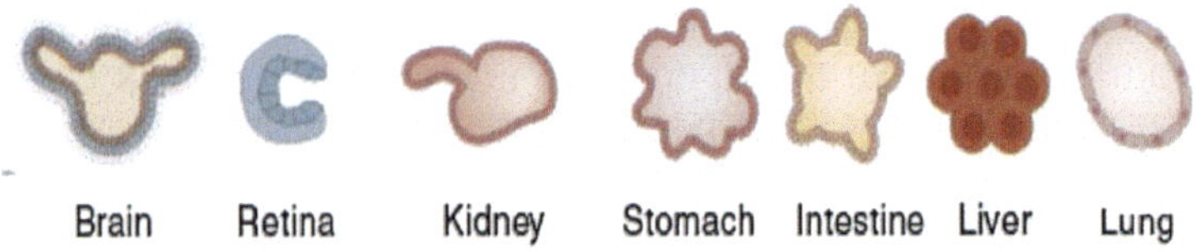

Fig. 10.1 Organoid examples

10.2.2 Latest Organoid Books and Their Main Ideas

Investigation from [58] highlights recent and emerging advances that describe *organoid differentiation protocols* for the different organ systems that implement organoids as tools to understand complexity and maturation, high content drug screening, disease modeling, development and evolution.

Brain organoid research is dicussed in [59], which explores *multiple methods and approaches used to generate human brain and neuroretinal organoids* to address fundamental questions in human brain research. Comprehensive and brain organoid research is a valuable resource for researchers at various levels of learning, ranging from undergraduate students to advanced laboratories.

Organoid bioengineering is investigated in [60], with organoids as three-dimensional miniature tissue established from embryonic stem cells, human pluripotent stem cells, adult stem cells, and cancer cells. These fascinating 3D organoids serve as a valuable tool for fundamental research, disease modeling, drug screening and discovery, regenerative medicine, and deciphering the mechanism of disease pathogenesis.

Organoid technology for disease modeling and personalized treatment is in [61]. It provides a comprehensive overview of current knowledge of the organoid as a human-organ-in-a-dish, a powerful new technology for studying fundamental aspects of human organ development and disease progression in the search for drugs for personalised treatment.

Stem cells, structure, and function of organoids are in [62], which addresses the challenge of how to instruct stem/early progenitor cells to progress through appropriate steps to generate functional 3-dimensional organs, one of the outstanding issues in regenerative medicine. The field of organoids is geared towards defining and demonstrating the in vitro conditions that achieve this goal.

Organoids and mini-organs are discussed in [63], which combines contributions from leading practitioners. It begins with an introduction, but then delves into chapters that present advice on how to make organoids for many systems. Case studies that illustrate the uses of organioids are presented, along with discussions on future directions and specific problems that need to be solved.

Tumor organoids are discussed in [64]. Cancer cell biology research in general, and anti-cancer drug development specifically, still relies on standard cell culture techniques that place the cells in an unnatural environment. By encapsulating the respective cell type or combining several cell types to form tissues, these model organs are cultured to develop functional properties similar to native tissues.

From organoid towards cancer tissue engineering is considerd in [65]. It focuses on fundamental and applied research involving the combination of biomaterials and cancer cells to develop a three-dimensional (3D) tumor microenvironment in vitro, in which carcinogenesis mechanisms can be studied and therapies can be screened. The aim is acceleration of more precise therapies to fight cancer. *Ethical and legal aspects of organoids* are considered in [66]. These are arising from the use of brain organoids for research, therapeutic and enhancement purposes. Until recently, only post-mortem tissue was available for a structural examination of the brain. However, in order to better understand the development and function of the human brain, dynamic and functional investigations of different human brain cells are necessary.

10.2.3 Examples of Investigated Cells-Based Techniques

- **Cell differentiation and tissue**

Essentials of cell biology are considered in [67]. Within multicellular organisms, tissues are organized communities of cells that work together to carry out a specific function, hey also have different transcriptional programs and may well divide at different rates, Without cell division, long-term tissue survival would be impossible. Transcription factors can turn on at different times during cell differentiation, see Fig. 10.2.

- **Shapes of organoids**

Tissue geometry driving organoid patterning is in [68]. Organoids can mimic real organs, but guiding stem cells to grow an organoid of defined shape and size is difficult. Bioengineers have developed new methods for successfully guiding the stem cells to grow into intestinal tissues with real-life 3D structure and function. Examples of possible organoid shapes shown in Fig. 10.3.

The *extracellular matrix and cell wall* are considered in [69]. Plants and fungi have a tough cell wall for protection and support, while animal cells can secrete materials into their surroundings to form a meshwork of macromolecules called the extracellular matrix. The paper looks in more detail at these external structures and the roles they play in different cell types. The extracellular matrix is directly connected to the cells it surrounds.

- **Worms**

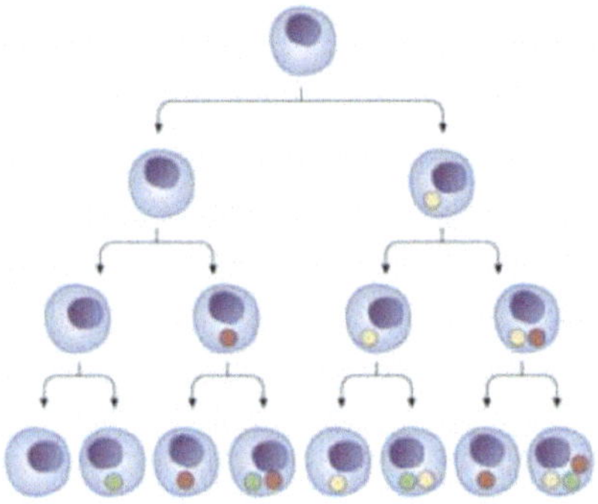

Fig. 10.2 Cells division and proliferation

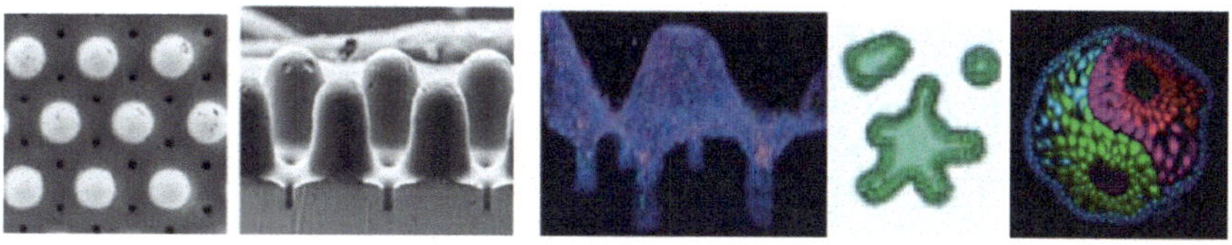

Fig. 10.3 Different organoid shapes

Fig. 10.4 Worms

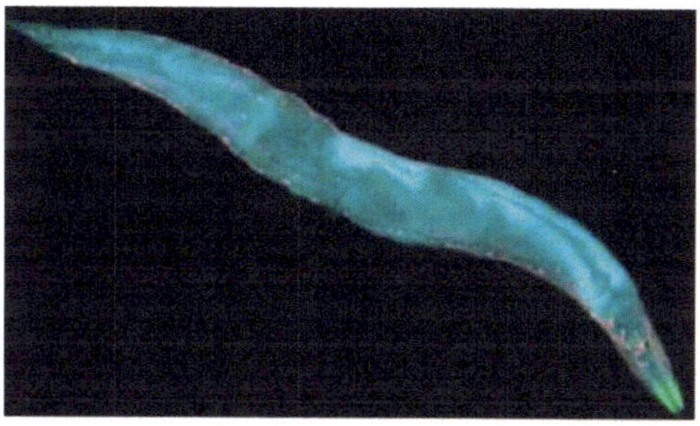

Cell migration in worms in relation to human cancer is in [70]. A new study has used migrating stem cells in flat worms to investigate regenerative properties that are found in cancer cells. Researchers have taken advantage of a special type of worm that is full of stem cells that migrate. The worms, called planarians, are known for their ability to regenerate all their tissues and organs repeatedly again and again.

What worms can tell us about brains and behaviour is discussed in [71]. Published in the prestigious journal Cell, the MIT team produced an atlas of how each brain cell encodes the animal's actions, revealing the underlying "logic" of how the worm's brain produces a sophisticated and flexible repertoire of behaviors. See also Fig. 10.4.

- **Killer cells**

Killer Cells and their behavior are discussed in [72]. Natural Killer (NK) Cells are lymphocytes in the same family as T and B cells, coming from a common progenitor. However, as cells of the innate immune system, NK cells respond quickly to a wide variety of pathological challenges. NK cells are best known for killing virally infected cells and detecting and controlling early signs of cancer. See Fig. 10.5.

10.3 Basics of Spatial Grasp Model and Technology

Only main technology issues are mentioned here, with details in Chaps. 3, 4 and 5, more in [73–91]. Within Spatial Grasp Model and Technology (SGT), a high-level operational scenario expressed in recursive Spatial Grasp Language (SGL), starting in any world point *propagates, covers, and matches the distributed environment in*

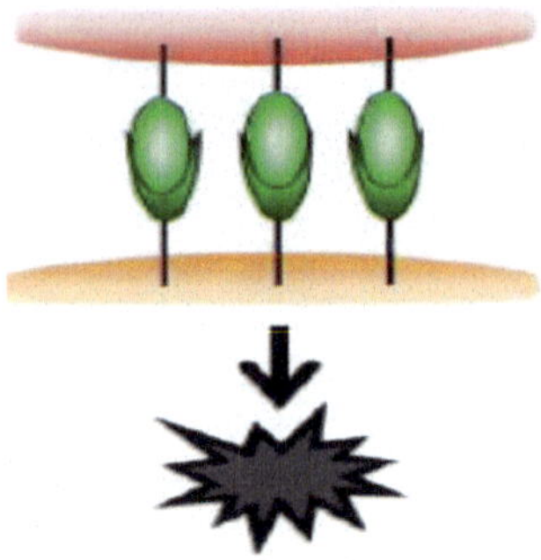

Fig. 10.5 Killer cells operation

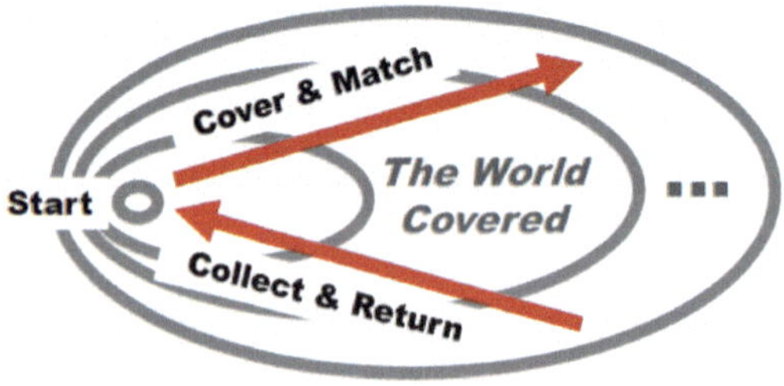

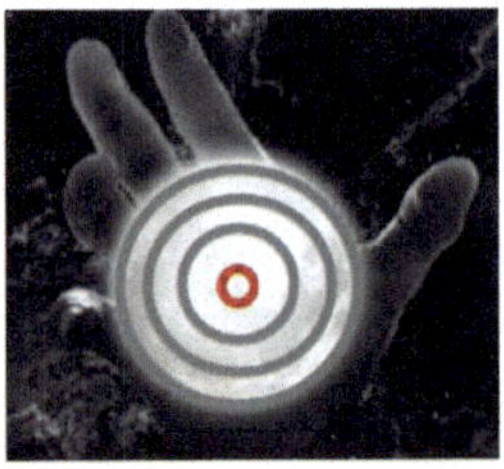

Fig. 10.6 Parallel wavelike world coverage under spatial grasp model

parallel wavelike mode as symbolically shown in Fig. 10.6. Such propagation can result in returning and analyzing the reached states and data, which may be arbitrarily remote, or used for launching more waves.

The SGL allows for organizing *direct space presence and operations* with unlimited powers and parallelism. Its universal recursive organization, with operational scenarios called *grasp*, can be expressed by a single formula:

$$\textbf{\textit{grasp}} \quad \rightarrow \textbf{\textit{constant}}|\textbf{\textit{variable}}|\textbf{\textit{rule}}\ (\{\textbf{\textit{grasp}},\})$$

The *rule* expresses certain action, control, description or context accompanied with operands, which can be any *grasps* too. The rules, starting in certain points, can organize navigation of the world sequentially, in parallel, or any combinations thereof. They can result in the same application points or cause movement to other world points with obtained results left there or returned.

The *SGL interpreter* consists of specialized modules serving SGL scenarios or their parts happened to be inside this interpreter, also organizing exchanges with other interpreters for distributed SGL scenarios. Communicating SGL interpreters can be in arbitrary number of copies, representing *powerful spatial engines operating without central resources or control*, as in Fig. 10.7. They can effectively work with *spatial graph and network data* of any volumes and distributions.

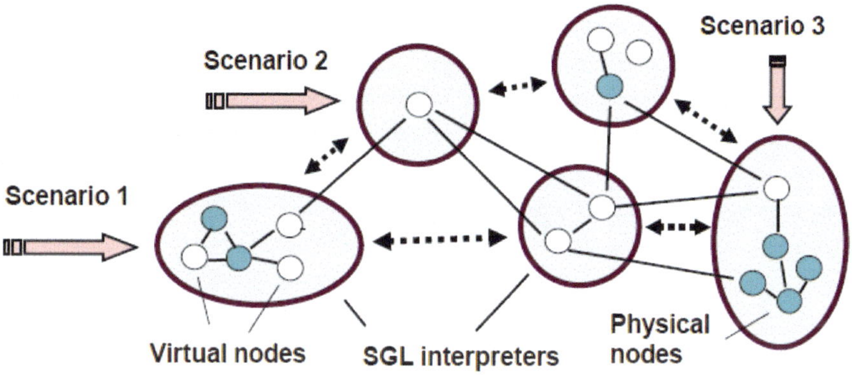

Fig. 10.7 SGL distributed networked interpretation

10.4 Examples of Networking Under SGT

Showing here how some elementary operations on distributed networks can be expressed in SGL.

- **Creation, movement, reshaping of nodes**

Creation of a node named a initially in a virtual space, changing its size, then moving into physical space with proper coordinates, and subsequent reshaping is shown in Fig. 10.8, also by the SGL expression below.

```
create_node(a); SIZE = bigger; WHERE = x1_y1;
move(x2_y2); SHAPE = triangle
```

- **Creation of networks in virtual and physical spaces**

Different variants of creation of networks are shown in Fig. 10.9.

In Fig. 10.9a and the following SGL scenario a virtual tree-like network is created in parallel top-down mode, having certain names of nodes (a, b, c, d, e, f, g) and connecting them links (all named as l).

```
create(node(a);
     ((link(l), node(b)); (link(l), nodes(d, e))),
```

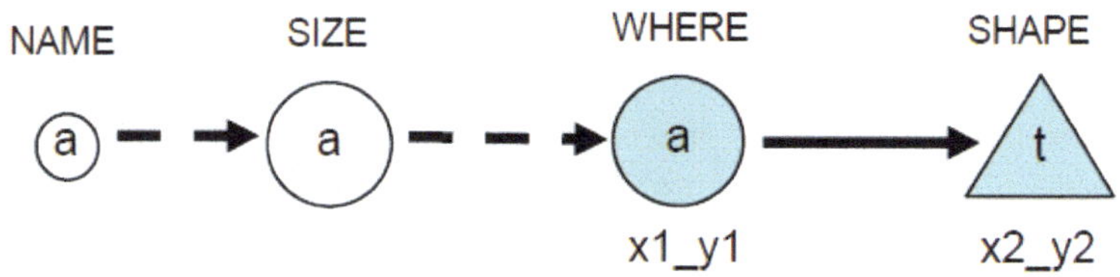

Fig. 10.8 Node creation and moving

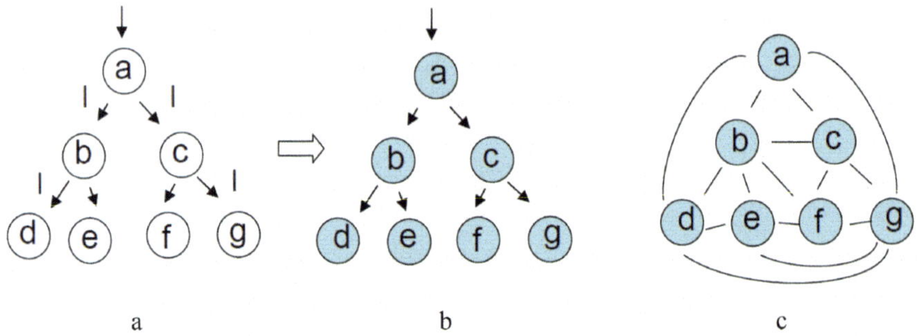

Fig. 10.9 Network creation

```
((link(l), node(c)); (link(l), nodes(f, g))))
```

In Fig. 10.9b and the following scenario, the creation of a similar network (with same structure and names of nodes and links) but also covering physical space (with physical coordinates of nodes x as Wx) is shown.

```
create(node(a_Wa);
     ((link(l), node(b_Wb));
      (link(l), nodes(d_Wd, e_We))),
     ((link(l), node(c_Wc));
      (link(l), nodes(f_Wf, g_Wg))))
```

For Fig. 10.9c with same nodes and link names but more complex network structure, the following scenario shows how to create this whole network in parallel, using its advance Top description (keeping node names and physical addresses, also names of direct neighbors of each node), which first splits by its nodes, then synchronously creates all these nodes in parallel. After this each node forms links to the already existing neighboring nodes (solving competition which neighbor node of the two should create this link).

```
Top = (a_Wa:(b,c,d,g), b_Wb:(a,c,d,e,f),
     c_Wc:(a,b,f,g), d_Wd(a,b,e,g),
     e_We:(b,d,f,g), f_Wf:(e,b,c,g),
     g_Wg:(a,c,f,d,e))
align(split(Top); frontal(Next) = VAL[2];
    create(VAL[1]));
hop(Next); NAME > BEFORE; linkup(1, BEFORE)
```

- **Finding paths in the network**

It is easy to find different paths in networks in SGL. For example, all simple paths from node a to node g (i.e. the ones with non repeating nodes) can be easily found by:

```
frontal(Path); hop(a);
repeat(
```

Fig. 10.10 Finding other nodes in a physical space from a certain node

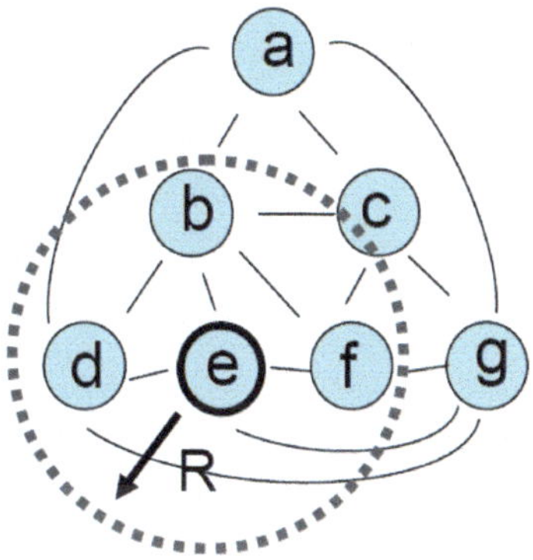


```
append(Path, NAME);
if(NAME == g, done(output(Path)));
  hop(all_links); notbelong(NAME, Path))
```

If we apply this scenario to the network of Fig. 10.9a, b, will receive a single path only, like: (a,c,g). But if applied to the more complex network of Fig. 10.9c, we will receive multiple simple paths, as follows:

```
(a,c,g),(a,c,f,g),(a,b,c,f,g),(a,b,c,g),(a,b,d,e,f,g),
(a,b,e,f,g),(a,b,e,g),(a,b,f,g)
```

- **Finding proper nodes in virtual and physical spaces**

It is possible to address network nodes in very different ways: by their names, addresses, by links to or from other nodes, their individual physical locations, belongings to some territory or area defined properly, or by physical distances from other nodes. For example, doing this from certain node e and distance R from it, we may receive the following nodes, see also Fig. 10.10:

For: `output(hop(e); hop(distance < R))`, will have `(d, b, f)`.

For: `output(hop(e); hop(distance > R))`, will have `(a, c, g)`.

10.5 Organoid Models Under SGT

We will consider here examples of how SGL spatial networking can express and simulate some basic organoid features.

10.5.1 Cells Growth, Division and Replication

Such activities are shown in Fig. 10.11 in relation to those already discussed in [67].

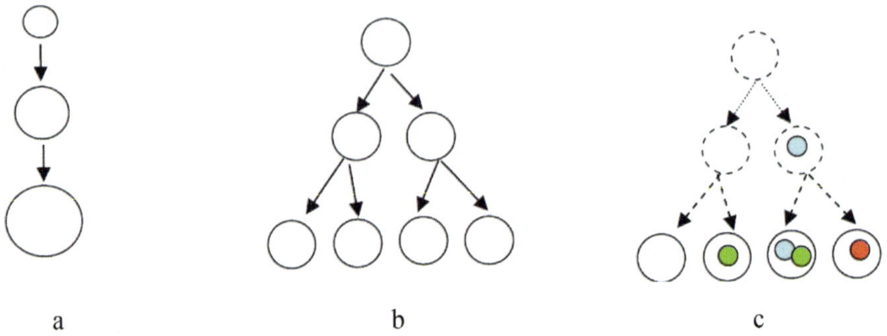

Fig. 10.11 Cells dynamics: **a** growth; **b** replication; **c** using transcriptional regulators

Just staying in the same place and increasing its size in time can be expressed as follows, see Fig. 10.11a.

```
create_node(x);
repeat(steps)(increase(SIZE); sleep(time))
```

Abstract cells replication scheme in time can be expressed as (also Fig. 10.11b):

```
hop(x); repeat(steps)(create(double(x)); sleep(time))
```

More realistic scenario for the cells replication may be when their doubling result includes the original nodes (i.e. the node logically providing replication symbolically disappears as a unit afterwards), is in physical space with certain distances between cells, and allows new cells to obtain additional features (say, expressed by different transcriptional colors), as below, and in Fig. 10.11c.

```
frontal(Color = (w,b,g,r), R = radius);
hop(x); WHERE = start;
repeat(steps)(
  sequence(
    create(double(x, random(Color), distance(R))),
    remove(current));
  sleep(time))
```

10.5.2 Worm Creation—Moving

One of the living sells possibilities is their collective movement in physical space as a worm, as discussed in [70, 71]. Figure 10.12 first shows the creation of original virtual worm structure, and then, starting from the first cell, moving in physical space with pulling behind the following cells, which stepwise obtain physical existence and finally become the completed worm. The following SGL scenario describes this worm creating process with the whole worm moving in the same direction, each time using the same incremental position change by the first cell.

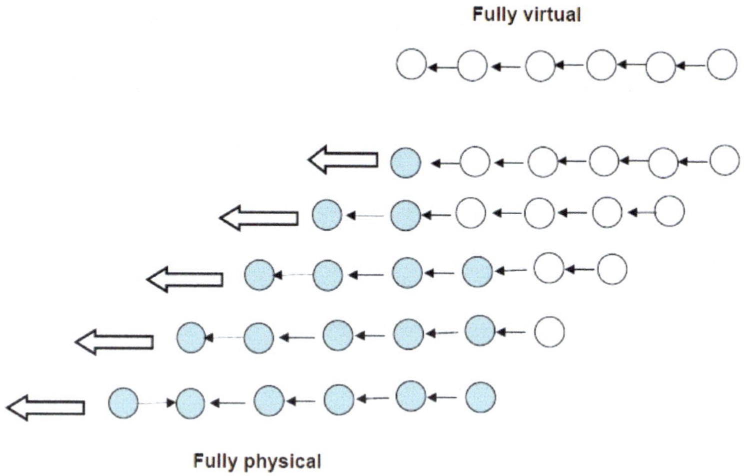

Fig. 10.12 Worm creation and movement

```
frontal(Shift = dX_dY, Start = X_Y, Old, old1);
create_node(x);
stay(repeat(five)(create(link(+1), node(x))));
WHERE = Start;
repeat(
 Old = WHERE; SHIFT(Shift);
 stay(repeat(hop(+1); Old1 = WHERE; WHERE = Old;
        Old = Old1)));
```

The worm creation and movement in physical space via any route, as in Fig. 10.13, may be organized in SGL as follows. For this, the head cell instead of same incremental move is supplied with full set of route coordinates from the beginning, taking the next one each time and dragging behind the remaining cells of the worm. Each remaining cell will be following the position of the previous cell during their collective movement, reflecting altogether the established curvature of the route.

```
Points = coordintes; Start = X_Y; frontal(Old, old1);
create_node(x);
stay(repeat(five)(create(link(+1), node(x))));
WHERE = Start;
repeat(
 Old = WHERE; WHERE = withdraw(Points, 1);
 stay(repeat(hop(+1); Old1 = WHERE; WHERE = Old;
```

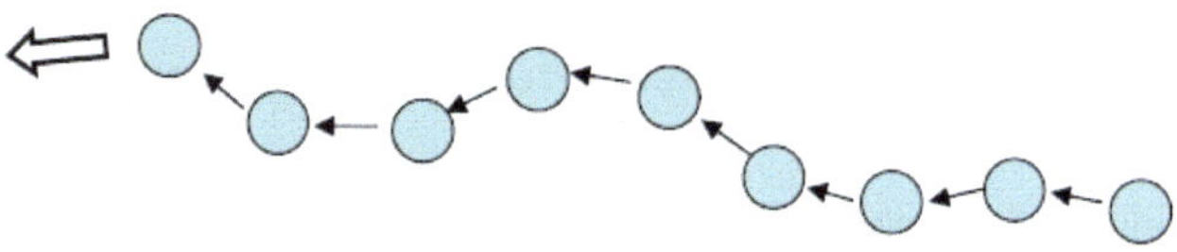

Fig. 10.13 Worm following any path

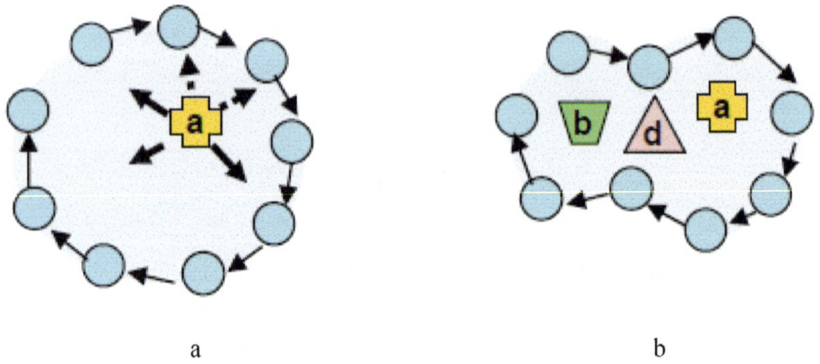

Fig. 10.14 Possible forming of organoid boundaries by a worm

```
Old = Old1)))
```

This simulated worm technique forming curved lines of interconnected cells can be used for different purposes. For example, potentially reflecting any shapes (very different shapes discussed in [68, 69]), it can effectively outline the whole regions to be occupied by organoids, like in Fig. 10.14. This will allow cells to move only inside and not outside the region, as in Fig. 10.14a, also permitting multiple cells to grow, replicate, move and communicate exclusively within its borders, as in Fig. 10.14b. Of course, such organoid borders can also be created in SGL at once and in parallel, if all their coordinates are planned in advance.

10.5.3 Killer Cell Operation

Natural Killer (or NK) cells, as in [72], are propagating, detecting, and killing infected cells, including those related to cancer. It is easy to represent this activity in SGL, see also Fig. 10.15, where the killer node (in red) is propagating in physical space and searching for and removing cancerous cells reached on some distance, as follows.

```
stay(create_number(node(cancer, random(area(x1_y1))));
 create(node(killer, x2_y2));
 repeat(move_random(R);
     if(see(danger, distance), kill_remove))
```

10.5.4 Simplified Full Organoid Model and Its Operation

Having considered above and tested some elementary techniques for dealing with cell models and their interactions, we can now compose a very simple united organoid

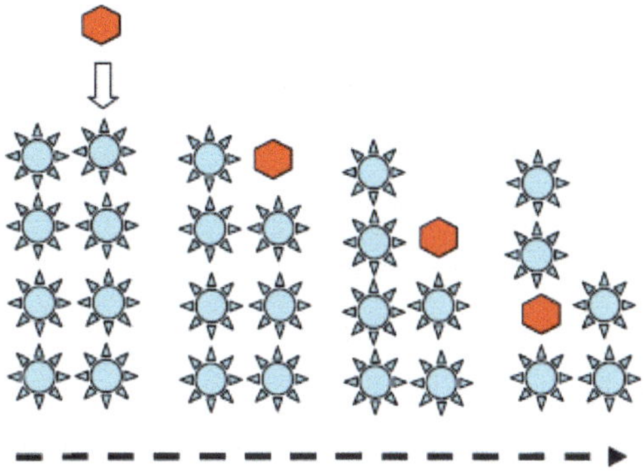

Fig. 10.15 Killer cell operation

model, which can autonomously exist and evolve in physical space. This is symbolically shown in Fig. 10.16a for the initial organoid stage, and in Fig. 10.16b for some further stages in time, as well as by its implementation with the following active SGL code. This scenario first creates organoid borders, and then inhabits this region with different types of cell nodes named `a`, `b`, `c`, `d` (including cancer cells `c`), which may self-grow and self-replicate taking into account their individual features, if this is allowed by still remaining available space inside organoid borders. The scenario also injects into the organoid space a number of killer nodes (each named `k`), which will be randomly propagating inside the organoid space and looking for cancer `c` nodes, while killing and removing them when seen at the proper distance. The released physical spaces after removal of cancer nodes may be immediately and effectively used by other cell types for their further growth and replication, and so on.

```
Border = (x1_y1, x2_y2, ..., xn_yn);
Types = (a, b, c, d);
stay(create_nodes(Border));
parallel(
  (create_nodes(Types, random(inside(Border)));
   repeat(
    move(style(NAME), random(R), inside(Border));
```

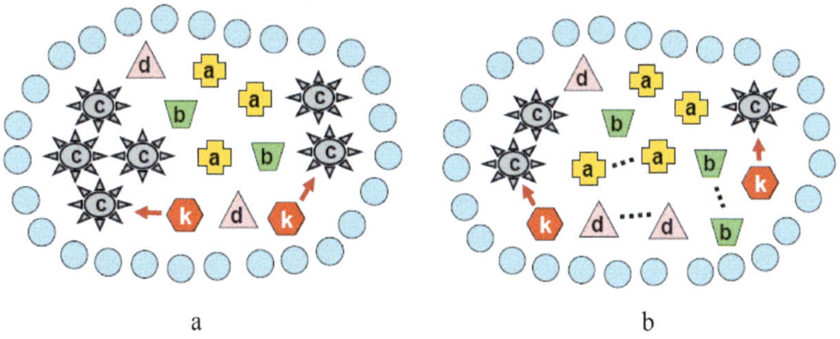

Fig. 10.16 Simplified model of organoid

```
    if(empty(around(R), inside(Border)),
      replicate(myself));
    sleep(delay1))),
  (create_number(k, random(inside(Border)));
   repeat(
    move(style(NAME), random(R), inside(Border));
    if(see(Types, R), kill_remove);
    sleep(delay2))))
```

10.5.5 Investigating Parameters of Growing Organoids

The previously described simple organoid model may be further investigated during its operation in time, using different overall parameters hinting on the state and quality of its evolution in time, like the following.

Counting the current number of all healthy cells:

```
output(count(hop(a,b,d)))
```

Counting the current number of all still existing cancer cells:

```
output(count(hop(c)))
```

Counting the current proportion of cancer cells to all other cells, reflecting in some sense the overall "health" of this organoid, which may vary in time due to different individual cell growth and evolution, also deletion of killed cancer nodes which can self-grow and self-replicate too:

```
output(count(hop(c))/count(hop(a,b,d)))
```

10.6 Other Organoid Developments of Interest

Informing here some about some newest organoid features and capabilities, which may be of particular interest for simulation and practical use by the described networking technology, to be analyzed in detail and reported in subsequent reports and publications.

- **Robotics**

Connecting laboratory grown "mini brains" to robots is considered in [92]. A cerebral organoid is hooked up to a computer that's also linked to a spiderlike, four-legged robot. The computer picks up spontaneous electrical signals from the organoid, and then, based on programming from the researchers, feeds information to the robot after which it walks forward.

Tiny living robots made from human cells can move, as in [93]. Scientists have created tiny living robots from human cells that can move around in a lab dish and may one day be able to help heal wounds or damaged tissue, according to a new study. They moved in different ways and survived up to 60 days in laboratory conditions.

Having sufficient experience of using networking models for individual and collective behavior of advanced robots, SGT can also simulate the creation and behavior of such biological brains, effectively integrating them with powerful control and management solutions in SGL, thus contributing to the development of advanced physical-virtual robotic brains with practically unlimited applications.

- **Brain waves**

Cerebral organoids produce complex brain waves similar to newborns, as in [94]. Despite their small size, the cerebral organoids generated several types of brain waves, according to electrodes placed at multiple spots on them. Their neurons fired at frequencies indistinguishable from those that animate actual brains, including gamma waves, alpha waves, and delta waves. No one has ever seen this level of complexity in cerebral organoids,

Grown mini brain in a lab has human-like brain waves, as in [95]**.** Does the brain organoid have a consciousness? This work really shows that the organoid has complex patterns of neural activity for future studies. Growing mini brains on-demand, ones with actual brain waves, could enable researchers to rapidly carry out trials and test drugs that might lead to better treatments.

Fundamentally based on self-spreading recursive waves which can create any distributed networks, with the latter spreading waves in distributed spaces themselves, SGT can effectively simulate the creation and behavior of such waves-producing organoids, moving together with these biological mini-brains towards simulation and understanding of this great but mysterious phenomenon called *consciousness* [82, 86].

- **Mini brains in space**

Growing mini human brains in space is discussed in [96]. Mini-sized, round brain organoids are being developed. Organoids from stem cells in the skin are grown in outer space. These organoids can be used for neurological disease therapy, drug testing, and biocomputer development. The report that brain organoids are unlikely to have the ability to "think" in the future as feared.

Time traveling mini-brains to conquer space are in [97]. It discusses the launch of brain organoids into outer space and how microgravity enriches our understanding of brain development and disease. In terms of cellular aging, one month on the space station is equivalent to between ten to thirty years on earth. The idea of using the space station as an incubator for aging may be transformative. This great idea of moving organoid creation and investigation into outer apace may have enormous applications, where simulation and creation of living tissue beyond Earth may be on the way to conquest of Universe. The SGT potentially uniting any terrestrial and celestial networking facilities for arbitrary complex spatial solutions, can be used in full for this mission (including networking and integration of biological and virtual brains from different space locations).

10.7 Conclusions

As the result of this chapter we can acknowledge the tested capability and efficiency of using the developed Spatial Grasp networking paradigm to express and simulate practically any organization, behavior and functionality of individual and multiple cells with their combinations and interactions up to the whole organoids, traditionally grown in vitro mode in laboratories. We have shown some very basic organoid simulation technique in SGL, with plans of more detailed and broad investigations and results to be obtained. Such results can believably be used in different cases, as follows.

First, the virtual networking models of organoids can be obtained much quicker than the growing organoid tissue in laboratories, and the resultant advanced networking solutions, which can employ any computational facilities throughout the whole internet, can be quickly and in detail analyzed for different, like medical, purposes.

Second, if living tissue organoids are really needed, like in medical surgery and organ replacement, then advanced virtual modeling of the growth of such organoids by SGT can assist, properly orient, and coordinate the development of the related living tissue.

Third, it may appear useful to strategically combine virtual an live growth and investigation in the united organoid development process, where living tissue growth can be under constant supervision and control by the intelligent SGT networks, the latter dynamically evolving in the combined virtual-physical space too.

References

1. Transport networks. Inspire Registry, (2015). https://inspire.ec.europa.eu/theme/tn
2. Transportation networks. Travel Forecasting Resource. https://tfresource.org/topics/Transportation_networks.html
3. Durkin, J., Goodman, J., et al.: Building multiservice transport networks (networking technology), 1st edn. Cisco Press, (2012)
4. Murakami, M., Li, H., Ryoo, J-D.: Packet transport networks: overview and future direction. In: 2 nd APT/ITU conformance and interoperability workshop (C&I-2), (2014). https://www.itu.int/en/ITU-T/C-I/interop/Documents/20140826/CI-2_INP-09_Packet_Transport_Networks_Overview_and_Future_Direction.pdf
5. Air Transport Network. https://en.wikipedia.org/wiki/Air_Transport_Network
6. Qureshi, K.N., Ul Islam, F., et al.: Improved road segment-based geographical routing protocol for vehicular ad-hoc networks. Electronics **9**(8), 1248 (2020). https://doi.org/10.3390/electronics9081248. https://www.mdpi.com/2079-9292/9/8/1248
7. Sujan.: What is communication network? Definition, importance, and 4 types (explained), (2023). https://tyonote.com/communication_network/

8. Fitzek, F.H.P., Granelli, F., et al.: Computing in communication networks: from theory to practice, 1st edn. Academic Press, (2020)
9. Srikant, R., Ying, L.: Communication networks: an optimization, control, and stochastic networks perspective, 1st edn. Cambridge University Press, (2013)
10. Raghunandan, K.: Introduction to wireless communications and networks: a practical perspective (Textbooks in telecommunication engineering), 1st edn, 2022 edition. Springer, (2023)
11. Lou, Y., Ahmed, N.: Underwater communications and networks (Textbooks in telecommunication engineering). Springer, (2021)
12. Kaufman, C., Perlman, R., et al.: Network security: private communication in a public world (Prentice hall series in computer networking and distributed systems), 3rd edn. Addison-Wesley Professional, (2022)
13. Wu, Y., Singh, S., et al.: 6G mobile wireless networks (Computer communications and networks), 1st edn, 2021 edn. Springer, (2022)
14. Social network. https://en.wikipedia.org/wiki/Social_network
15. Kenton, W.: What is social networking? Investopedia, Updated May 24, 2023. https://www.investopedia.com/terms/s/social-networking.asp#:~:text=Social%20networking%20refers%20to%20using,)%2C%20Instagram%2C%20and%20Pinterest
16. Nanawati, S.: Social network analytics. Published in Analytics Vidhya, (2019). https://medium.com/analytics-vidhya/social-network-analytics-f082f4e21b16
17. Singh, C.: 20+ Most popular social networking sites in 2024. (2023). https://www.socialpilot.co/social-networking-sites
18. Jackson, M.O.: Social and economic networks, Illustrated edn. Princeton University Press, (2010)
19. Harrison,T.: Battle networks and the future force. Center for Strategic and International Studies. (2021). https://www.csis.org/analysis/battle-networks-and-future-force
20. Stillion, J., Clark, B.: What it takes to win: succeeding in 21st century battlenetwork conmpetitions, Center for Strategic and Budgetary Assessments (CSBA), (2015). https://csbaonline.org/uploads/documents/What-it-Takes-to-Win.pdf
21. Hadley, G.: DAF outlines a new 'Battle network' as its contribution to JADC2. Air & Space Forces Mag., (2023). https://www.airandspaceforces.com/daf-battle-network-contribution-jadc2/
22. The challenges of military networks and technological developments, elbit systems. https://elbitsystems.com/e-lynx_multi_channel/
23. Themnér, A., Karlén, N.: Building a safety net: explaining the strength of ex-military networks. Secur. Stud., **29**(2), 268–300. https://doi.org/10.1080/09636412.2020.1722851. https://www.tandfonline.com/doi/full/10.1080/09636412.2020.1722851
24. What is an economic network? Advantages and disadvantages of economic networks, Vietnambiz, (2019). https://vietnambiz.vn/mang-luoi-kinh-te-economic-network-la-gi-uu-nhuoc-diem-cua-mang-luoi-kinh-te-20191011113428212.htm
25. Halton, C.: Economic network: What it is, pros and cons, examples. Updated October 24, 2022. https://www.investopedia.com/terms/e/economic-network.asp#:~:text=Key%20Takeaways-,An%20economic%20network%20is%20a%20combination%20of%20individuals%2C%20groups%2C%20or,companies%20or%20partnerships%20between%20corporations
26. Todo, Y., Matous, P.: Economic networks for more innovative and resilient economies. https://cepr.org/voxeu/columns/economic-networks-more-innovative-and-resilient-economies
27. Emmert-Streib, F., Tripathi, S., et al.: Understanding the world economy in terms of networks: a survey of data-based network science approaches on economic networks. Front. Appl. Math. Stat., (2018). Sec. Math. Financ., **4**, (2018). https://doi.org/10.3389/fams.2018.00037. https://www.frontiersin.org/articles/10.3389/fams.2018.00037/full
28. Wixted, B.: Innovation system frontiers: cluster networks and global value (Advances in spatial science). Springer, (2009)
29. Manyika, J.: What the economy looks like so far for individuals in the 21st century. (2020). https://www.linkedin.com/pulse/what-economy-looks-like-so-far-individuals-21st-century-james-manyika

30. Hayes, A.: Distribution network: definition, how it works, and examples, Investopedia. Updated February 22, 2021. https://www.investopedia.com/terms/d/distribution-network.asp#:~:text=In%20a%20supply%20chain%2C%20a,or%20through%20a%20retail%20network
31. Distribution Network.: The flow of goods from a producer or supplier to an end consumer. Written by CFI Team. https://corporatefinanceinstitute.com/resources/valuation/distribution-network/
32. What are types of distribution networks in supply chain management? oboloo FAQ's. https://oboloo.com/blog/what-are-types-of-distribution-networks-in-supply-chain-management/
33. Distribution network—What is it and what are the benefits? https://www.salesbabu.com/blog/what-is-a-distribution-network-and-its-benefits/
34. Wu, Q., Shen, F., et al.: Optimal operation of active distribution networks: congestion management. Academic Press, Voltage Control and Service Restoration (2023)
35. Watts, S.: What is a virtual network? BMC, (2020). https://www.bmc.com/blogs/virtual-network/
36. What is Virtual networking? VMware by Broadcom. https://www.vmware.com/topics/glossary/content/virtual-networking.html
37. Network virtualization. https://en.wikipedia.org/wiki/Network_virtualization
38. Carlos, O., Duarte, M.B., Pujolle, G.: Virtual networks: pluralistic approach for the next generation of internet (Networks and telecommunications), 1st edn. Wiley-ISTE, (2013)
39. Fundamentals of virtual networking. https://www.geeksforgeeks.org/fundamentals-of-virtual-networking/
40. Neural network. https://en.wikipedia.org/wiki/Neural_network
41. Anand, V.: Introduction to neural network in machine learning. Updated On August 10th, 2023. https://www.analyticsvidhya.com/blog/2022/01/introduction-to-neural-networks/
42. Ashtari, H.: What is a neural network? Definition, working, types, and applications in 2022. Last Updated: August 3, 2022. https://www.spiceworks.com/tech/artificial-intelligence/articles/what-is-a-neural-network/
43. Artificial Neural Network. NVIDIA Developer. https://developer.nvidia.com/discover/artificial-neural-network
44. Wu, L., Cui, P., et al.: Graph neural networks: foundations, frontiers, and applications, 1st edn, 2022 edn. Springer, (2022)
45. Borsboom, D., Deserno, M.K., Rhemtulla, M. et al.: Network analysis of multivariate data in psychological science. Nat Rev Methods Primers. **1**, 58 (2021). https://doi.org/10.1038/s43586-021-00055-w. https://www.nature.com/articles/s43586-021-00055-w
46. Jones, P.J., Mair, P., McNally, R.J.: Visualizing psychological networks: a tutorial in R. Front. Psychol, Sec. Quant. Psychol. Meas., **9**, (2018). https://doi.org/10.3389/fpsyg.2018.01742. https://www.frontiersin.org/articles/10.3389/fpsyg.2018.01742/full
47. Fonseca-Pedrero, E.: Network aanalysis in psychology. Psychol. Pap., **39**(1), 1–12 (2018). https://doi.org/10.23923/pap.psicol2018.2852. https://www.papelesdelpsicologo.es/pdf/2852.pdf
48. Bringmann, L.F.: Person-specific networks in psychopathology: Past, present, and future. Curr. Opin. Psychol., **41**, 59–64 (2021). https://www.sciencedirect.com/science/article/pii/S2352250X21000300?via%3Dihub
49. Yuqing, C., Shuyang, D., et al.: Network analysis and its applications in psychology. Adv Psychol Sci, **28**(1), 178–190 (2020). https://doi.org/10.3724/SP.J.1042.2020.00178, https://journal.psych.ac.cn/adps/EN/https://doi.org/10.3724/SP.J.1042.2020.00178
50. Campana, P.: Explaining criminal networks: Strategies and potential pitfalls. Methodol. Innov. **9**, 1–10 (2016). https://doi.org/10.1177/2059799115622748. https://journals.sagepub.com/doi/pdf/https://doi.org/10.1177/2059799115622748
51. Bichler, G.: Understanding criminal networks: a research guide, 1st edn. University of California Press, (2019)
52. Morselli, C.: Crime and networks (Criminology and justice studies). Routledge, (2013)
53. New forms of organized crime: networked structure. UNODS, Unite Nations Office on Drugs and Crime. https://www.unodc.org/e4j/en/organized-crime/module-7/key-issues/networked-structure.html

54. Hufnagel, S., Moiseienko, A.: Criminal networks and law enforcement: global perspectives on illegal enterprise (Transnational criminal justice), 1st edn. Routledge, (2019)
55. Detecting and disrupting transnational criminal organizations: analytics for interdependent smuggling and money-laundering networks, criminal investigation and network analysis, DHS centers of excellence. https://cina.gmu.edu/projects/detecting-and-disrupting-transnational-criminal-organizations-analytics-for-interdependent-smuggling-and-money-laundering-networks/
56. Organoid. https://en.wikipedia.org/wiki/Organoid
57. What are organoids? https://www.mblbio.com/bio/g/product/cell-culture/article/organoid_overview.html
58. Spence, J.: Human pluripotent stem cell derived organoid models (ISSN book 159), 1st edn. Academic Press, (2020)
59. Gopalakrishnan, J.: Brain organoid research (Neuromethods book 189). Humana, (2022)
60. Paul, M.K.: Organoid bioengineering—advances, applications and challenges (biomedical engineering). IntechOpen, (2022)
61. Yahaya (ed.).: Organoid technology for disease modelling and personalized treatment (Stem cell biology and regenerative medicine book 71). Humana, (2022)
62. Turksen, K.: Organoids: stem cells, structure, and function (Methods in molecular biology, 1576), 1st ed. Springer, (2019)
63. Davies, J.A., Lawrence, M.: Organoids and mini-organs, 1st edn. Academic Press, (2018)
64. Soker, S., Skardal, A.: Tumor organoids (Cancer drug discovery and development), 1st ed, 2018 edition. Humana, (2017)
65. Danti, S., Negrini, N.C., et al.: Biomaterial-assisted 3d in vitro tumor models: from organoid towards cancer tissue engineering approaches. Mdpi AG, (2023)
66. Dederer, H.-G., Hamburger, D.: Brain organoids in research and therapy: fundamental ethical and legal aspects (Advances in neuroethics), 1st ed, 2022 edn. Springer, (2022)
67. Essentials of cell biology, Unit 5: How do cells know when to divide? https://www.nature.com/scitable/ebooks/essentials-of-cell-biology-14749010/118241827/
68. Gjorevski, N., Nikolaev, M., et al.: Tissue geometry drives deterministic organoid patterning. Science, (2022). https://www.ncbi.nlm.nih.gov/pmc/articles/PMC9131435/
69. The extracellular matrix and cell wall. AP.BIO, ENE-2 (EU), ENE-2.D (LO), ENE-2.D.1 (EK), ENE-2.D.2 (EK). https://www.khanacademy.org/science/ap-biology/cell-structure-and-function/membrane-permeability/a/the-extracellular-matrix-and-cell-wall
70. Keeping moving—stem cell migration in worms is controlled by genes involved in human cancer. Oxford, Department of biology. https://www.biology.ox.ac.uk/article/keeping-moving-stem-cell-migration-in-worms-is-controlled-by-genes-involved-in-human-cancer
71. What worms can tell us about brains and behaviour. Science news, (2023). https://www.leeds.ac.uk/news-science/news/article/5384/what-worms-can-tell-us-about-brains-and-behaviour#:~:text=Researchers%20at%20Massachusetts%20Institute%20of,insights%20about%20its%20nervous%20system
72. Eissmann, P.: Natural killer cells. ImperialCollege, London, UK. https://www.immunology.org/public-information/bitesized-immunology/cells/natural-killer-cells
73. Sapaty, P.S.: A distributed processing system. European Patent N 0389655, Publ. 10.11.93, European Patent Office
74. Sapaty, P.S.: Mobile processing in distributed and open environments. John Wiley & Sons, New York (1999)
75. Sapaty, P.S.: Ruling distributed dynamic worlds. John Wiley & Sons, New York (2005)
76. Sapaty, P.S.: Managing distributed dynamic systems with spatial grasp technology. Springer, (2017)
77. Sapaty, P.S.: Holistic analysis and management of distributed social systems. Springer, (2018)
78. Sapaty, P.S.: Complexity in international security: a holistic spatial approach. Emerald Publishing, (2019)
79. Sapaty, P.S.: Symbiosis of real and simulated worlds under spatial grasp technology. Springer, (2021)

80. Sapaty, P.S.: Spatial grasp as a model for space-based control and management systems. CRC Press, (2022)
81. Sapaty, P.S.: The spatial grasp model: applications and investigations of distributed dynamic worlds. Emerald Publishing, (2023)
82. Sapaty, P.S.: Providing integrity, awareness, and consciousness in distributed dynamic systems. CRC Press, (2024)
83. Sapaty, P.S.: Spatial management of air and missile defence operations. Math. Mach. Syst. (1), (2023). http://www.immsp.kiev.ua/publications/articles/2023/2023_1/01_23_Sapaty.pdf
84. Sapaty, P.S.: Providing distributed system integrity under spatial grasp technology. Math. Mach. Syst. 2, (2023). http://www.immsp.kiev.ua/publications/articles/2023/2023_2/02_23_Sapaty.pdf
85. Sapaty, P.S.: Providing global awareness in distributed dynamic systems. Int. Relat. Dipl., **11**(2), 87–100 (2023). https://doi.org/10.17265/2328-2134/2023.02.002. https://www.davidpublisher.com/Public/uploads/Contribute/6486c3d05a6cc.pdf
86. Sapaty, P.S.: Simulating distributed consciousness with spatial grasp model. Math. Mach. Syst. (3), (2023)
87. Sapaty, P.S.: Network centrality operations under spatial grasp technology. J. Adv. Artif. Intell. Mach. Learn., **1**(1), 1–11 (2023). https://www.scivisionpub.com/pdfs/network-centrality-operations-under-spatial-grasp-technology-3025.pdf
88. Sapaty, P.S.: Managing distributed systems with spatial grasp patterns. Math. Mach. Syst. (4), (2023)
89. Sapaty, P.S.: Basic network operations under spatial grasp model. Sci Set J Phys. 3(1), 01–16 (2024). https://mkscienceset.com/articles_file/500-_article1708322845.pdf
90. Sapaty, P.S.: Spatial networks as models for organoid cultures and brain research. Math. Mach. Syst. (1), (2024)
91. Sapaty, P.S.: Basic network operations under spatial grasp model. Sci Set J of Physics **3**(1), 01–16 (2024)
92. Schley, L.: Meet the scientists connecting lab-grown "Mini Brains" to robots, discover. (2019). https://www.discovermagazine.com/mind/meet-the-scientists-connecting-lab-grown-mini-brains-to-robots
93. Hunt, K.: Tiny living robots made from human cells surprise scientists. CNN, (2023). https://edition.cnn.com/2023/11/30/world/living-robots-from-human-cells-scn/index.html
94. Begley.: In a first, cerebral organoids produce complex brain waves similar to newborns', reviving ethical concerns, STAT. (2019). https://www.statnews.com/2019/08/29/in-a-first-cerebral-organoids-produce-complex-brain-waves-similar-to-newborns-reviving-ethical-concerns/
95. Carey, T.: Scientists grew a mini brain in a lab. It has human-like brain waves. What does that mean for research? Freethink. (2019). https://www.freethink.com/science/lab-grown-mini-brain
96. Wahyudi, M.Z.: Growing mini human brains in space. (2023). https://www.kompas.id/baca/english/2023/08/01/en-menumbuhkan-otak-mini-manusia-di-luar-angkasa
97. Kulbatski, I.: Time traveling mini-brains on a missionto conquer space. (2022). https://www.the-scientist.com/news-opinion/time-traveling-mini-brains-on-a-mission-to-conquer-space-70353

Chapter 11
Conclusions

11.1 Introduction

The book reviewed, analyzed and classified many existing publications on the use of networks in very different areas. It described the developed universal spatial paradigm capable to create, support, integrate and manage any networking systems and provide effective solutions in them. The chapter is organized as follows:

Section 11.2 repeats different types of networks reviewed, investigated and analyzed with the help of SGT. Section 11.3 briefs basics of Spatial Grasp Model, Technology and Language used in the book. Section 11.4 reveals selected main points expressing and confirming suitability and efficiency of SGT and SGL for global networking and management of large distributed systems. Section 11.5 hypothesizes a possible three-level vision of the distributed word, with Spatial Grasp philosophy and model providing global awareness and consciousness at the top, multi-networking level materialized under SGT at the middle, and real physical, virtual and executive world at the bottom, under management from the networking level. Section 11.6, as Final Notes, includes technology implementation possibilities and future plans for the research in progress. References cite numerous sources discovered, analyzed, classified, and used for practical solutions in previous chapters of the book.

11.2 The Sources Reviewed, Analyzed, and Used in the Book

These include the following.

P. S. Sapaty, *Spatial Networking in the United Physical, Virtual, and Mental World*, Studies in Systems, Decision and Control 533,
https://doi.org/10.1007/978-3-031-62154-3_11

- *Other books on the networking topic* [1–17]. Which includes some other books published in this and related areas, to learn how they are organized, their volume, when published, how may relate to the current book proposal, and so on.
- *Networking in different areas* [9, 18–71]. Networks as models and operations are widely used in practically every area of human activity. Contains analyzed and reviewed very different types and areas of networking.
- *Network Analysis* [14, 72–78]. Reviews and analyses existing publications on network analysis, where network is analyzed as a collection of interconnected entities or nodes that can communicate or exchange information with one another
- *Network Management* [8, 79–85]. Reviews and analyses existing publications on network management, which refers to the process of administering and monitoring the network to keep things running smoothly and properly.
- *Basic Network operations* [86–98]. These sources were analyzed and investigated in detail for drafting the repertoire of main networking operations which may be useful, even critical, for their usage within areas described in the book.
- *Network centrality* [99–105]. Contains summary on existing network centrality publications, reflecting network centrality, centrality in social networks, centrality and underground network, centrality in psychological networks, centrality in transportation network design, and others.
- *Combined production-delivery networks* [106–114]. This includes papers on Production Networks, Product Evaluation Networks, and Product Distribution Networks, where the latter have to interconnect the consumption and production sides.
- *On Patterns* [115–141]. Where patterns may relate to almost everything around us. It includes papers on pattern theory, different pattern types, pattern matching and recognition, and pattern languages.
- *On Organoids* [142–158]. Where organoid is miniaturized and simplified version of an organ produced in vitro that mimics the key functional, structural and biological complexity of that organ.

11.3 Spatial Grasp Technology as the Book's Basic Philosophy and Method

The book has been strongly influenced by the Spatial Grasp paradigm, technology and language summarized below, also detailed in Chaps. 3, 4 and 5, European patent [159], previous books [160–168], latest papers [169–176], many other publications too, with older ones relating to previous versions called WAVE and mobile agents [177–187]. Within Spatial Grasp Model and Technology (SGT), a high-level operational scenario expressed in recursive Spatial Grasp Language (SGL), starting in any world point *propagates, covers, and matches the distributed environment in parallel wavelike mode.* As symbolically shown in Fig. 11.1, Such propagation can result in returning and analyzing the reached states and data, which may be arbitrarily remote, or used for launching more waves.

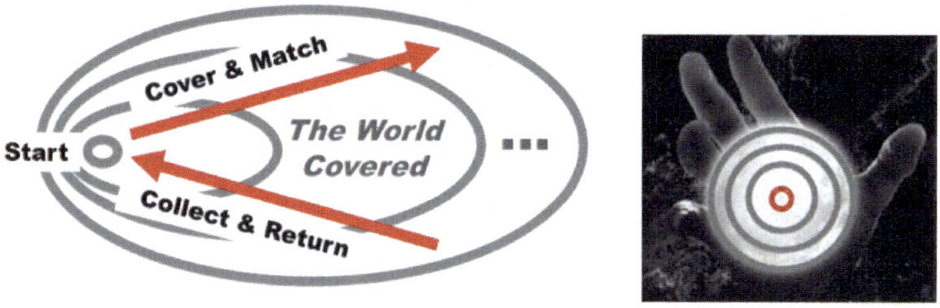

Fig. 11.1 Parallel wavelike world coverage under spatial grasp model

The SGL allows for *direct space presence and operations* with unlimited powers and parallelism. Its universal recursive organization, with operational scenarios called *grasp*, can be expressed by a single formula:

$$\boldsymbol{grasp} \quad \to \boldsymbol{constant}|\boldsymbol{variable}|\boldsymbol{rule}\ (\{\boldsymbol{grasp},\})$$

The *rule* expresses certain action, control, description or context accompanied with operands, which can be any *grasps* too. The rules, starting in certain points, can organize navigation of the world sequentially, in parallel, or any combinations thereof. They can result in the same application points or cause movement to other world points with obtained results left there or returned.

The *SGL interpreter* consists of specialized modules serving SGL scenarios or their parts happened to be inside this interpreter, also organizing exchanges with other interpreters for distributed SGL scenarios. Communicating SGL interpreters can be in arbitrary number of copies, representing *powerful spatial engines operating without central resources or control*, as in Fig. 11.2. They can effectively work with *spatial graph and network data* of any volumes, distributions, and coverage.

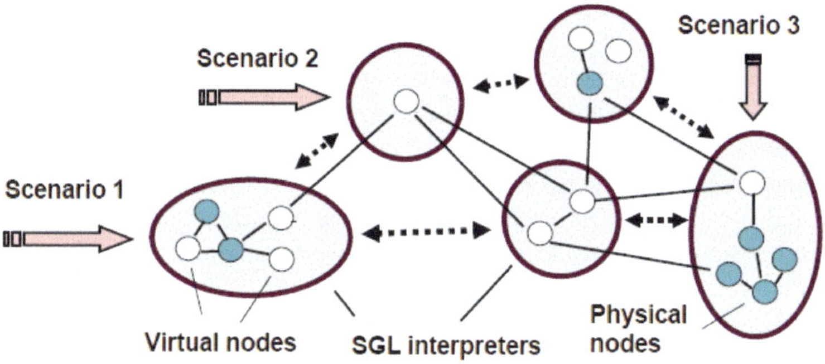

Fig. 11.2 SGL distributed networked interpretation

11.4 Suitability of SGT and SGL for Global Networking

After the detailed analysis of the use of networks in many publications, also gained practical experience in using SGT and SGL for solving networking problems in very different areas an their integration, as in previous chapters, we share the following conclusions.

(1) The results of this book confirm *suitability and efficiency of SGT and SGL* in development of networking methods for solving complex problems within combined distributed systems.
(2) They can work with *arbitrary complex and large networks* distributed worldwide, throughout any terrestrial and celestial environments.
(3) These networks can *reflect physical, virtual, command, control*, and any united systems and solutions used in civil and defense areas.
(4) The networks implemented or simulated under SGT can be *highly dynamic, both predictable and unpredictable* if needed, naturally reflecting real world situations.
(5) These resultant networks *can also become active and self-organizing*, behaving as autonomous management and control systems *superior* to the systems they oversee, being, in their turn, *overseen by Spatial Grasp philosophy and paradigm* at the highest level.
(6) The technology allows us to obtain *detailed, very clear and extremely compact* solutions for solving the network problems, which can operate on arbitrary large and complex networks *in highly parallel and fully distributed* mode.
(7) These solutions operate *without vulnerable central resources*, at the same time providing high integrity, awareness and even consciousness of different technical, virtual, biological, social or combined systems.
(8) Expressed in high-level recursive SGL, the implementation scenarios are *extremely compact* in comparison with other languages, being based on *holistic spatial pattern recognition and matching* rather than traditional algorithmic cultures.
(9) The investigated paradigm allows us to *directly move and operate* in physical, virtual and mental spaces by clear and compact mode, radically *differing from traditional algorithmic thinking.*
(10) Deep integration of networked heterogeneous distributed systems can be organized *naturally and automatically* within the *global integrity, awareness, and consciousness* processes under Spatial Grasp paradigm.

11.5 Integral Heterogeneous Multi-Networking World Under SGT

We have analyzed and reviewed very different areas of networking, including transport, communication, social, battle, military, economic, distribution, virtual, neural, psychological, criminal, production, evaluation, and distribution; algorithms and operations on graphs networks; also many graph and networking solutions and applications investigated and implemented under Spatial Grasp paradigm.

After detailed analysis and provided networking solutions, we may expect the developed Spatial Grasp Technology and its basic Spatial Grasp Language to be considered as a theoretical and practical concept and tool for working with any types of networks. By the results obtained in this book and also reflected in numerous publications on the this Spatial Grasp paradigm in social, civil, security, defense, and many other areas, all with obtained networking solutions, the whole world can be symbolically considered as integration of the following levels, see also Fig. 11.3:

Level 3: Holistic spiritual and mental level providing and supporting integrity, awareness and consciousness for the whole system, as in [163, 165, 168, 170, 172, 176], Chap. 10 too.

Level 2: Networking models and concepts which can effectively reflect and control large, heterogeneous and complex world (as in Chaps. 2, 6, 7, 8 and 9).

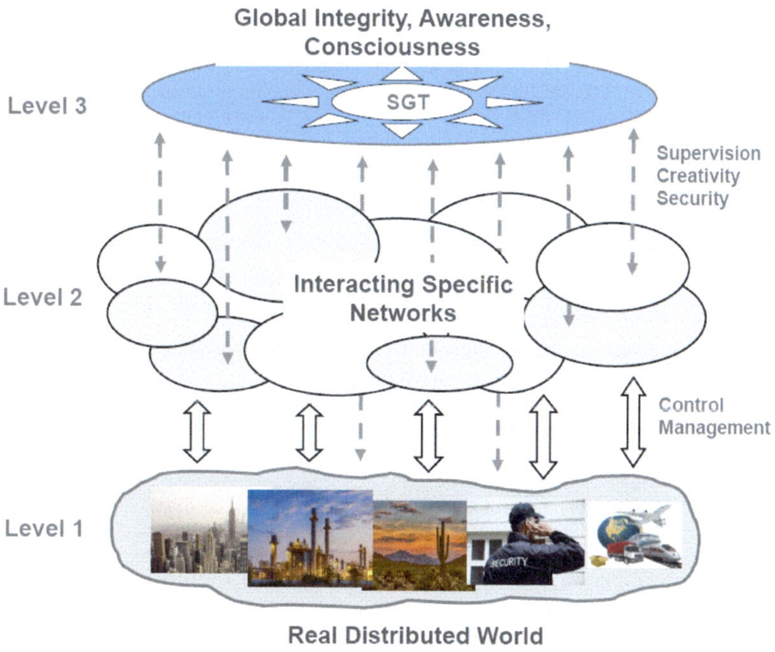

Fig. 11.3 Hypothetical three-level world organization under SGT

Level 1: Real world integrating physical, virtual and executive components, which may have any terrestrial and celestial nature and overage (as classified in Chap. 3).

These levels can operate together as follows. Highest Level 3 describes, materializes, and supervises Level 2 which, in turn, controls and manages Level 1. Level 3 can also, if needed, directly access Level 1 to analyze its state and needs and design, create, or update proper networking models for Level 2, which, in turn, should govern Level 1 with the delegated power and reasonable autonomy from Level 3.

11.6 Final Notes

The latest version of SGT *can be implemented* by a small group of system programmers, even in standard company or university environments, as was done for the previous versions in different countries with the author's supervision. *Future plans* of this work include application of SGT and SGL for extended network-based systems in numerous areas, including economy, ecology, management, security, international relations, education, philosophy, psychology, psychiatry, and many others.

References

1. Wright, D.L.S.: The physical and virtual space of the consulting room: room-object spaces. Routledge, (2022)
2. Weiss, P.L., Keshner, E.A., et al.: Virtual reality for physical and motor rehabilitation (Virtual reality technologies for health and clinical applications). Springer, (2014)
3. Friedman, P.: The pilates method of physical and mental conditioning. Doubleday, (1984)
4. Wixted, B.: Innovation system frontiers: cluster networks and global value (Advances in spatial science). Springer, (2009)
5. Morsink, R.L.A.: Foreign direct investment and corporate networking: a framework for spatial analysis of investment conditions. Edward Elgar Publishing, (1999)
6. Bai, L., Liang, X., et al.: Spatial multidimensional cooperative transmission theories and key technologies. World Scientific, (2020)
7. Yang, K.S., Shekhar, S.: Spatial network big databases: queries and storage methods. Springer, (2017)
8. Franke, M.: Managing airline networks: Design. Routledge, Integration and innovative technologies (managing aviation operations) (2020)
9. Wu, Q., Shen, F., et al.: Optimal operation of active distribution networks: congestion management. Academic Press, Voltage control and service restoration (2023)
10. Isvoranu, A.-M., Epskamp, S., et al.: Network psychometrics with r: a guide for behavioral and social scientists. Routledge, (2022)
11. Meghanathan, N.: Centrality metrics for complex network analysis: emerging research and opportunities (Advances in wireless technologies and telecommunication). IGI Global, (2018). Lange, A.: Centrality in strategic transportation network design: An application to less-than-truckload networks (Edition KWV). Springer Gabler, (2019)
12. Morselli, A.: Crime and networks (Criminology and Justice Studies). Routledge, (2013)
13. Read, J., Bentall, R., et al.: Models of madness: psychological, social and biological approaches to psychosis (The international society for psychological and social approaches to psychosis book series). Routledge, (2013)

14. Diwadkar, V.A., Eickhoff, S.B.: Brain network dysfunction in neuropsychiatric illness: methods, applications, and implications. Springer, (2021)
15. Spence, J.: Human pluripotent stem cell derived organoid models (ISSN Book 159), 1st edn. Academic Press, (2020)
16. Gopalakrishnan, J.: Brain organoid research (Neuromethods book 189). Humana, (2022)
17. Paul, M.K.: Organoid bioengineering—advances, applications and challenges (Biomedical engineering). IntechOpen, (2022)
18. Transport networks.: Inspire Registry. (2015). https://inspire.ec.europa.eu/theme/tn
19. Transportation networks, Travel Forecasting Resource. https://tfresource.org/topics/Transportation_networks.html
20. Durkin, J., Goodman, J., et al.: Building multiservice transport networks (Networking technology), 1st edn. Cisco Press, (2012)
21. Murakami, M., Li, H., Ryoo, J-D.: Packet transport networks: overview and future direction. In: 2nd APT/ITU conformance and interoperability workshop (C&I-2). (2014). https://www.itu.int/en/ITU-T/C-I/interop/Documents/20140826/CI-2_INP-09_Packet_Transport_Networks_Overview_and_Future_Direction.pdf
22. Air Transport Network. https://en.wikipedia.org/wiki/Air_Transport_Network
23. Qureshi, K.N., Ul Islam, F., et al.: Improved road segment-based geographical routing protocol for vehicular Ad-hoc networks. Electronics **9**(8), 1248 (2020). https://doi.org/10.3390/electronics9081248. https://www.mdpi.com/2079-9292/9/8/1248
24. Sujan.: What is communication network? definition, importance, and 4 types (Explained), (2023). https://tyonote.com/communication_network/
25. Fitzek, F.H.P., Granelli, F., et al.: Computing in communication networks: from theory to practice, 1st edn. Academic Press, (2020)
26. Srikant, R., Ying, L.: Communication networks: an optimization, control, and stochastic networks perspective, 1st edn. Cambridge University Press, (2013)
27. Raghunandan, K.: Introduction to wireless communications and networks: a practical perspective (Textbooks in telecommunication engineering), 1st ed., 2022 edition. Springer, (2023)
28. Lou, Y., Ahmed, N.: Underwater communications and networks (Textbooks in telecommunication engineering). Springer, (2021)
29. Kaufman, C., Perlman, R., et al.: Network security: private communication in a public world (prentice hall series in computer networking and distributed systems), 3rd edn. Addison-Wesley Professional, (2022)
30. Wu, Y., Singh, S., et al.: 6G mobile wireless networks (computer communications and networks), 1st ed. 2021 edition. Springer, (2022)
31. Social network. https://en.wikipedia.org/wiki/Social_network
32. Kenton, W.: What is social networking? Investopedia. Updated May 24, 2023. https://www.investopedia.com/terms/s/socialnetworking.asp#:~:text=Social%20networking%20refers%20to%20using,)%2C%20Instagram%2C%20and%20Pinterest
33. Nanawati, S.: Social network analytics, published in analytics vidhya. (2019). https://medium.com/analytics-vidhya/social-network-analytics-f082f4e21b16
34. Singh, C.: 20+ most popular social networking sites in 2024. (2023). https://www.socialpilot.co/social-networking-sites
35. Jackson, M.O.: Social and economic networks, Illustrated edition. Princeton University Press, (2010)
36. Harrison, T.: Battlenetworks and the future force, center for strategic and international studies. (2021). https://www.csis.org/analysis/battle-networks-and-future-force
37. Stillion, J., Clark, B.: What it takes to win: succeeding in 21st century battlenetwork competitions. Center for Strategic and Budgetary Assessments (CSBA), (2015). https://csbaonline.org/uploads/documents/What-it-Takes-to-Win.pdf
38. Hadley, G.: DAF outlines a new 'battle network' as its contribution to JADC2. Air & Space Forces Mag., (2023). https://www.airandspaceforces.com/daf-battle-network-contribution-jadc2/

39. The Challenges of military networks and technological developments, Elbit Systems. https://elbitsystems.com/e-lynx_multi_channel/
40. Themnér, A., Karlén, N.: Building a safety net: explaining the strength of ex-military networks. Secur. Stud., **29**(2), 268–300. https://doi.org/10.1080/09636412.2020.1722851. https://www.tandfonline.com/doi/full/10.1080/09636412.2020.1722851
41. What is an Economic Network? Advantages and disadvantages of economic networks. Vietnambiz, (2019). https://vietnambiz.vn/mang-luoi-kinh-te-economic-network-la-gi-uu-nhuoc-diem-cua-mang-luoi-kinh-te-20191011113428212.htm
42. Halton, C.: Economic network: what it is, pros and cons, examples. Updated October 24, 2022. https://www.investopedia.com/terms/e/economic-network.asp#:~:text=Key%20Takeaways-,An%20economic%20network%20is%20a%20combination%20of%20individuals%2C%20groups%2C%20or,companies%20or%20partnerships%20between%20corporations
43. Todo, Y., Matous, P.: Economic networks for more innovative and resilient economies. (2015). https://cepr.org/voxeu/columns/economic-networks-more-innovative-and-resilient-economies
44. Emmert-Streib, F., Tripathi, S., et al.: Understanding the world economy in terms of networks: a survey of data-based network science approaches on economic networks. Front. Appl. Math. Stat., Sec. Math. Financ., **4**, 2018. https://doi.org/10.3389/fams.2018.00037. https://www.frontiersin.org/articles/10.3389/fams.2018.00037/full
45. Wixted, B.: Innovation system frontiers: cluster networks and global value (Advances in spatial science). Springer, (2009)
46. Manyika, J.: What the economy looks like so far for individuals in the 21st century. (2020). https://www.linkedin.com/pulse/what-economy-looks-like-so-far-individuals-21st-century-james-manyika
47. Hayes, A.: Distribution network: definition, how it works, and examples, Investopedia. Updated February 22, 2021. https://www.investopedia.com/terms/d/distribution-network.asp#:~:text=In%20a%20supply%20chain%2C%20a,or%20through%20a%20retail%20network.
48. Distribution Network. The flow of goods from a producer or supplier to an end consumer. Written by CFI team. https://corporatefinanceinstitute.com/resources/valuation/distribution-network/
49. What are types of distribution networks in supply chain management? oboloo FAQ's. https://oboloo.com/blog/what-are-types-of-distribution-networks-in-supply-chain-management/
50. Distribution Network—What is it and what are the benefits? https://www.salesbabu.com/blog/what-is-a-distribution-network-and-its-benefits/
51. Watts, S.: What is a virtual network? BMC, (2020). https://www.bmc.com/blogs/virtual-network/
52. What is Virtual Networking? VMware by Broadcom. https://www.vmware.com/topics/glossary/content/virtual-networking.html
53. Network virtualization. https://en.wikipedia.org/wiki/Network_virtualization.
54. Carlos, O., Duarte, M.B., Pujolle, G.: Virtual networks: pluralistic approach for the next generation of internet (networks and telecommunications), 1st edn. Wiley-ISTE, (2013)
55. Fundamentals of Virtual Networking. https://www.geeksforgeeks.org/fundamentals-of-virtual-networking/
56. Neural network. https://en.wikipedia.org/wiki/Neural_network
57. Anand,V.: Introduction to neural network in machine learning, Updated On August 10th, 2023. https://www.analyticsvidhya.com/blog/2022/01/introduction-to-neural-networks/
58. Ashtari, H.: What is a neural network? definition, working, types, and applications in 2022. Last Updated: August 3, 2022. https://www.spiceworks.com/tech/artificial-intelligence/articles/what-is-a-neural-network/
59. Artificial Neural Network. NVIDIA Developer. https://developer.nvidia.com/discover/artificial-neural-network
60. Wu, L., Cui, P., et al.: Graph neural networks: foundations, frontiers, and applications, 1st ed, 2022 edition. Springer, (2022)

61. Borsboom, D., Deserno, M.K., Rhemtulla, M. et al.: Network analysis of multivariate data in psychological science. Nat Rev Methods Primers. **1**, 58(2021). https://doi.org/10.1038/s43586-021-00055-w. https://www.nature.com/articles/s43586-021-00055-w
62. Jones, P.J., Mair, P., McNally, R.J.: Visualizing psychological networks: a tutorial in R. Front. Psychol, Sec. Quant. Psychol. Meas., **9**, (2018). https://doi.org/10.3389/fpsyg.2018.01742. https://www.frontiersin.org/articles/10.3389/fpsyg.2018.01742/full
63. Fonseca-Pedrero, E.: Network analysis in psychology. Psychol. Pap., **39**(1), 1–12 (2018). https://doi.org/10.23923/pap.psicol2018.2852, https://www.papelesdelpsicologo.es/pdf/2852.pdf
64. Bringmann, L.F.: Person-specific networks in psychopathology: Past, present, and future. Curr. Opin. Psychol., **41**, 59–64 (2021). https://www.sciencedirect.com/science/article/pii/S2352250X21000300?via%3Dihub
65. Yuqing, C., Shuyang, D., et al.: Network analysis and its applications in psychology. Adv. Psychol. Sci. **28**(1), 178–190 (2020). https://doi.org/10.3724/SP.J.1042.2020.00178
66. Campana, P.: Explaining criminal networks: Strategies and potential pitfalls. Methodol. Innov. **9**, 1–10 (2016). https://doi.org/10.1177/2059799115622748. https://journals.sagepub.com/doi/pdf/10.1177/2059799115622748
67. Bichler, G.: Understanding criminal networks: a research guide, 1st edn. University of California Press, (2019)
68. Morselli, C.: Crime and networks (Criminology and justice studies). Routledge, (2013)
69. New forms of organized crime: networked structure. UNODS, Unite Nations Office on Drugs and Crime. https://www.unodc.org/e4j/en/organized-crime/module-7/key-issues/networked-structure.html
70. Hufnagel, S., Moiseienko, A.: Criminal networks and law enforcement: global perspectives on illegal enterprise (transnational criminal justice), 1st edn. Routledge, (2019).
71. Detecting and disrupting transnational criminal organizations: analytics for interdependent smuggling and money-laundering networks, criminal investigation and network analysis. DHS Centers of Excellence. https://cina.gmu.edu/projects/detecting-and-disrupting-transnational-criminal-organizations-analytics-for-interdependent-smuggling-and-money-laundering-networks/
72. What is Network Analysis—An overview, by Great Learning. Updated on Jul 14, 2023. https://www.mygreatlearning.com/blog/what-is-network-analysis/#:~:text=Network%20analysis%20is%20a%20method,network%20functions%20as%20a%20whole
73. Rawlings, C.M.: Network analysis (structural analysis in the social sciences). Cambridge University Press, (2023)
74. Transport network analysis. https://en.wikipedia.org/wiki/Transport_network_analysis
75. Lange, A.: Centrality in strategic transportation network design: an application to less-than-truckload networks (Edition KWV). Springer Gabler, (2019)
76. Isvoranu, A.-M., Epskamp, S., et al.: Network psychometrics with R: A guide for behavioral and social scientists. Routledge, (2022)
77. Yang, K.S., Shekhar, S.: Spatial network big databases: queries and storage methods. Springer, (2017)
78. Goldenberg, D.: Social network analysis: from graph theory to applications with python. Published in Towards Data Science, (2021). https://towardsdatascience.com/social-network-analysis-from-theory-to-applications-with-python-d12e9a34c2c7
79. Iwuozor, J., Bottorff, C., Hoeper, M.: What is network management? Definition and best practices. (2023). https://www.forbes.com/advisor/business/software/what-is-network-management/
80. What is Network Management? https://www.opentext.com/what-is/network-management#:~:text=Network%20management%20is%20the%20sum,users%20efficiently%2C%20effectively%20and%20quickly
81. What Is Network Management? CISCO. https://www.cisco.com/c/en/us/solutions/enterprise-networks/what-is-network-management.html

82. BasuMallick, C.: What is network management? Definition, key components, and best practices. Spiceworks, (2022). https://www.spiceworks.com/tech/networking/articles/what-is-network-management/
83. Shin, B.: A practical introduction to enterprise network and security management, 2nd edn. Auerbach Publications, (2021)
84. Wilson, J., Trembath, M.: Asset management for road networks. SAP, (2021). https://rizing.com/wp-content/uploads/2021/08/SAP-Asset-Management-for-road-networks.pdf
85. Buhrkuhl, C.: Network management: A guide and best practices. Published: April 7, 2021. https://www.office1.com/blog/network-management-a-guide-and-best-practices
86. An Introduction to Networks. https://mathinsight.org/network_introduction
87. The Degree Distribution of a Network. https://mathinsight.org/degree_distribution
88. Random Networks. https://mathinsight.org/random_network
89. Network Algorithms. https://ics.uci.edu/~goodrich/teach/cs165/notes/NetworkAlgs.pdf
90. Newman, M.E.J.: 10 Fundamental network algorithms: A discussion of some of the most important and fundamental algorithms for performing network calculations on a computer. (2010). https://doi.org/10.1093/acprof:oso/9780199206650.003.0010
91. Graph and Network Algorithms. https://ch.mathworks.com/help/matlab/graph-and-network-algorithms.html
92. Schmid, S., Mandal, P.S.: Distributed network algorithms (Lecture notes for GIAN course). Summer, (2016). https://schmiste.github.io/GIAN-Lecture-Notes-NetAlg.pdf
93. Tarjan, R.E.: Data structures and network algorithms. CBMS-NSF Regional conference series in applied mathematics. https://epubs.siam.org/doi/book/10.1137/1.9781611970265
94. Ravikiran, A.S.: What is graph in data structure & types of graph? (2023). https://www.simplilearn.com/tutorials/data-structure-tutorial/graphs-in-data-structure
95. Graph Operations. https://pages.cs.wisc.edu/~paton/readings/Old/fall01/GRAPH-OPS.html
96. Basic Graph Theory. https://learn.saylor.org/mod/page/view.php?id=27304
97. Pandey, K.: Graph data structure—Explained with examples. (2022). https://www.masaischool.com/blog/graph-data-structure-explained-with-examples/
98. Ford-Fulkerson Algorithm for Maximum Flow Problem. https://www.scaler.com/topics/data-structures/ford-fulkerson-algorithm-for-maximum-flow-problem/
99. Bhasin, J.: Graph analytics—Introduction and concepts of centrality, towards data science. (2019). https://towardsdatascience.com/graph-analytics-introduction-and-concepts-of-centrality-8f5543b55de3
100. Rodrigues, F.A.: Network centrality: an introduction. (2019). arXiv:1901.07901v1. https://arxiv.org/pdf/1901.07901.pdf
101. Disney, A.: Social network analysis 101: centrality measures explained, Cambridge Intelligence. (2020). https://cambridge-intelligence.com/keylines-faqs-social-network-analysis/
102. Mussone, L., Viseh, H., Notari, R.: Novel centrality measures and applications to underground networks. Phys. A: Stat. Mech. Its Appl., **589**(1), 126595 (2022). https://www.sciencedirect.com/science/article/abs/pii/S0378437121008633
103. Bringmann, L.F., Elmer, T., Epskamp, S., Krause, R.W., et al.: What do centrality measures measure in psychological networks? J. Abnorm. Psychol., **128**(8), 892–903 0021–843X/19/$12.00 (2019). https://doi.org/10.1037/abn0000446. https://www.apa.org/pubs/journals/features/abn-abn0000446.pdf
104. Meghanathan, N.: Centrality metrics for complex network analysis: emerging research and opportunities (advances in wireless technologies and telecommunication). IGI Global, (2018). https://www.igi-global.com/book/centrality-metrics-complex-network-analysis/181520#table-of-contents
105. Lange, A.: Centrality in strategic transportation network design: an application to less-than-truckload networks (Edition KWV). Springer Gabler, (2019)
106. Global production network. https://en.wikipedia.org/wiki/Global_production_network
107. Lanza, G., Ferdows, F., Kara, S., et al.: Global production networks: Design and operation. CIRP Annals, **68**(2), 2019. https://www.sciencedirect.com/science/article/abs/pii/S0007850619301659

108. Gonzaga, F.P.: Local centers and connectivity: the foundation of global production networks, 1919. https://www.linkedin.com/pulse/local-centers-connectivity-foundation-global-networks-filipe
109. Collins, S., Yassine, A.A, Borgatti, S.P.: Evaluating product development systems using network analysis. Syst. Eng. **12**(1), 55–68 (2009). https://doi.org/10.1002/sys.20108. https://www.researchgate.net/publication/227602841_Evaluating_Product_Development_systems_using_network_analysis
110. Lin, H., Deng, X., et al.: A study of sustainable product design evaluation based on the analytic hierarchy process and deep residual networks. Sustainability **15**(19), 14538 (2023). https://doi.org/10.3390/su151914538. https://www.mdpi.com/2071-1050/15/19/14538
111. Danilchenko, K., Segal, M., Vilenchik, D.: Opinion spam detection: a new approach using machine learning and network-based algorithms. (2022). arXiv:2205.13422v1 [cs.LG]. https://www.semanticscholar.org/reader/1a6b30f29e2c9eee9eb34586bd9a1f9e822a5962
112. Bajor, P., Horvath, A.: The role of decision-making parameters in constructing and re-engineering of distribution networks. https://www.researchgate.net/publication/23637165_The_role_of_decision-making_parameters_in_constructing_and_re-engineering_of_distribution_networks
113. What is value delivery network in marketing? SCISPACE. https://typeset.io/questions/what-is-value-delivery-network-in-marketing-4um9b4s3z1
114. Distribution Network.: The flow of goods from a producer or supplier to an end consumer, CFI Team. https://corporatefinanceinstitute.com/resources/valuation/distribution-network/
115. Pattern. https://en.wikipedia.org/wiki/Pattern
116. Grenander, U., Miller, M.: Pattern theory: From representation to inference. Oxford University Press, p. 596. (2007)
117. De Souza, F.D.M., Sarkar, S., Srivastava, A., Su, J.: Pattern theory-based interpretation of activities. In: 22nd International conf on pattern recognition. (2014). https://projet.liris.cnrs.fr/imagine/pub/proceedings/ICPR-2014/data/5209a106.pdf
118. Mumford, D.: Pattern theory: A unifying perspective. https://www.dam.brown.edu/people/mumford/vision/papers/1994c-96--PattThUnifyingPersp-NC.pdf
119. Pattern theory. https://en.wikipedia.org/wiki/Pattern_theory
120. Joshi, U.: Patterns of distributed systems. https://martinfowler.com/articles/patterns-of-distributed-systems/
121. Sankey, T.: Statistical descriptions of spatial patterns. Encyclopedia of GIS. Springer, Cham (2017). https://link.springer.com/referenceworkentry/10.1007/978-3-319-17885-1_1351. https://doi.org/10.1007/978-3-319-17885-1_1351
122. Keim, D.A., Kröger, P., Ankerst, M.: Recursive pattern: A technique for visualizing very large amounts of data. IEEE Xplore: Conf. Visualization., (1995). https://www.researchgate.net/publication/3618236_Recursive_pattern_A_technique_for_visualizing_very_large_amounts_of_data. https://doi.org/10.1109/VISUAL.1995.485140.
123. Goertzel, B.: The hidden pattern: a patternist philosophy of mind, p 424. Brown Walker Press, (2006)
124. Russell, N., ter Hofstede, A.H.M., van der Aalst, W.M.P., Mulyar, N.: Workflow control-flow patterns: a revised view, p 134. BPMcenter, (2006)
125. van der Aalst, W.M.P., der Hofstede, A.H.M., Kiepuszewski, B., Barros, A.P.: Workflow patterns. (2002). http://www.workflowpatterns.com/documentation/documents/wfs-pat-2002.pdf
126. Pattern matching. https://en.wikipedia.org/wiki/Pattern_matching
127. Cheng, J., Xu Yu, J., Ding, B., Yu, P.S., Wang, H.: Fast graph pattern matching. https://www.microsoft.com/en-us/research/wp-content/uploads/2016/02/icde08gsearch.pdf
128. Reza, T.: Pattern matching in massive metadata graphs at scale. University of British Columbia, (2019). https://open.library.ubc.ca/soa/cIRcle/collections/ubctheses/24/items/1.0387453.https://doi.org/10.14288/1.0387453
129. Pattern recognition. https://en.wikipedia.org/wiki/Pattern_recognition

130. De Walker, V. R.: Consciousness is pattern recognition, Computer Science>Artificial Intelligence, Copyright 2016. https://arxiv.org/abs/1605.03009
131. Pattern recognition (psychology). https://en.wikipedia.org/wiki/Pattern_recognition_(psychology)
132. Pattern language. https://en.wikipedia.org/wiki/Pattern_language
133. A pattern language: towns, buildings, construction, 1977 book. https://en.wikipedia.org/wiki/A_Pattern_Language
134. What is a pattern language? https://groupworksdeck.org/pattern-language
135. Alexander, C.H., Ishikawa, S., Silverstein, M., Jacobson, M., Fiksdahl-King, I., Angel, S.H.: A pattern language: towns, buildings, construction. Oxford University Press, p. 1171. https://www.amazon.com/Pattern-Language-Buildings-Construction-Environmental/dp/0195019199
136. Winn, T., Calder, P.: A pattern language for pattern language structure. In: Third Asian pacific conference on pattern languages of programs. KoalaPLoP, (2002). https://www.researchgate.net/publication/228944468_A_pattern_language_for_pattern_language_structure.
137. Hafiz, M., Adamczyk, P., Johnson, R.: Growing a pattern language (for Security). In: Proc of the ACM international symposium on new ideas, new paradigms, and reflections on programming and software. (2012). https://www.researchgate.net/publication/262211609_Growing_a_pattern_language_for_security. https://doi.org/10.1145/2384592.2384607
138. Iba, T.: Pattern languages as media for the creative society. J. Inf. Process. Management., **55**(12), (2013). https://www.researchgate.net/publication/255484746_Pattern_Languages_as_Media_for_the_Creative_Society. https://doi.org/10.1241/johokanri.55.865
139. Shimomukai, E., Nakamura, S., Iba, T.: Change making patterns. a pattern language for fostering social entrepreneurship, p. 112. CreativeShift Lab, (2015)
140. Van Welie, M., van der Veer, G.: Pattern languages in interaction design: structure and organization. (2011). https://www.researchgate.net/publication/228881522_Pattern_languages_in_interaction_design_Structure_and_organization
141. Meszaros, G., Doble, J.: A pattern language for pattern writing. The Hillside Group. https://hillside.net/index.php/a-pattern-language-for-pattern-writing
142. Organoid.: https://en.wikipedia.org/wiki/Organoid
143. What are Organoids? https://www.mblbio.com/bio/g/product/cell-culture/article/organoid_overview.html
144. Spence, J.: Human pluripotent stem cell derived organoid models (ISSN book 159), 1st edn. Academic Press, (2020)
145. Gopalakrishnan, J.: Brain organoid research (Neuromethods book 189). Humana, (2022)
146. Paul, M.K.: Organoid bioengineering—advances, applications and challenges (Biomedical engineering). IntechOpen, (2022)
147. Yahaya, B.H., (ed).: Organoid technology for disease modelling and personalized treatment (Stem cell biology and regenerative medicine book 71). Humana, (2022)
148. Turksen, K.: Organoids: stem cells, structure, and function (methods in molecular biology, 1576), 1st ed. Springer, (2019)
149. Davies, J.A., Lawrence, M.: Organoids and mini-organs, 1st edn. Academic Press, (2018)
150. Soker, S., Skardal, A.: Tumor organoids (Cancer drug discovery and development), 1st ed., 2018 edition. Humana, (2017)
151. Danti, S., Negrini, N.C., et al.: Biomaterial-assisted 3D in vitro tumor models: from organoid towards cancer tissue engineering approaches. Mdpi AG, (2023)
152. Dederer, H.-G., Hamburger, D.: Brain organoids in research and therapy: fundamental ethical and legal aspects (Advances in neuroethics), 1st ed., 2022 edition. Springer, (2022)
153. Essentials of Cell Biology. Unit 5: How Do Cells Know When to Divide? https://www.nature.com/scitable/ebooks/essentials-of-cell-biology-14749010/118241827/
154. Gjorevski, N., Nikolaev, M., et al.: Tissue geometry drives deterministic organoid patterning. Science, (2022). https://www.ncbi.nlm.nih.gov/pmc/articles/PMC9131435/
155. The extracellular matrix and cell wall, AP.BIO, ENE-2 (EU), ENE-2.D (LO). ENE-2.D.1 (EK), ENE-2.D.2 (EK). https://www.khanacademy.org/science/ap-biology/cell-structure-and-function/membrane-permeability/a/the-extracellular-matrix-and-cell-wall

156. Keeping moving—stem cell migration in worms is controlled by genes involved in human cancer. Oxford, Department of biology. https://www.biology.ox.ac.uk/article/keeping-moving-stem-cell-migration-in-worms-is-controlled-by-genes-involved-in-human-cancer
157. What worms can tell us about brains and behaviour. Science news, (2023). https://www.leeds.ac.uk/news-science/news/article/5384/what-worms-can-tell-us-about-brains-and-behaviour#:~:text=Researchers%20at%20Massachusetts%20Institute%20of,insights%20about%20its%20nervous%20system.
158. Eissmann, P.: Natural killer cells. ImperialCollege, London, UK. https://www.immunology.org/public-information/bitesized-immunology/cells/natural-killer-cells
159. Sapaty, P.S.: A distributed processing system. European Patent N 0389655, Publ. 10.11.93, European Patent Office
160. Sapaty, P.S.: Mobile processing in distributed and open environments. John Wiley & Sons, New York (1999)
161. Sapaty, P.S.: Ruling distributed dynamic worlds. John Wiley & Sons, New York (2005)
162. Sapaty, P.S.: Managing distributed dynamic systems with spatial grasp technology. Springer, (2017)
163. Sapaty, P.S.: Holistic analysis and management of distributed social systems. Springer, (2018)
164. Sapaty, P.S.: Complexity in international security: a holistic spatial approach. Emerald Publishing, (2019)
165. Sapaty, P.S.: Symbiosis of real and simulated worlds under spatial grasp technology. Springer, (2021)
166. Sapaty, P.S.: Spatial grasp as a model for space-based control and management systems. CRC Press, (2022)
167. Sapaty, P.S.: The spatial grasp model: applications and investigations of distributed dynamic worlds. Emerald Publishing, (2023)
168. Sapaty, P.S.: Providing integrity, awareness, and consciousness in distributed dynamic systems. CRC Press, (2024)
169. Sapaty P.S.: Spatial management of air and missile defence operations. Math. Mach. Syst.. (1), (2023). http://www.immsp.kiev.ua/publications/articles/2023/2023_1/01_23_Sapaty.pdf
170. Sapaty, P.S.: Providing distributed system integrity under spatial grasp technology. Math. Mach. Syst..(2), (2023). http://www.immsp.kiev.ua/publications/articles/2023/2023_2/02_23_Sapaty.pdf
171. Sapaty P.S.: Providing global awareness in distributed dynamic systems. Int. Relat. Dipl., **11**(2), 87–100 (2023). https://doi.org/10.17265/2328-2134/2023.02.002. https://www.davidpublisher.com/Public/uploads/Contribute/6486c3d05a6cc.pdf
172. Sapaty P.S.: Simulating distributed consciousness with spatial grasp model. Math. Mach. Syst. (3), (2023)
173. Sapaty P.S.: Network centrality operations under spatial grasp technology. J. Adv. Artif. Intell. Mach. Learn., **1**(1), 1–11 (2023). https://www.scivisionpub.com/pdfs/network-centrality-operations-under-spatial-grasp-technology-3025.pdf
174. Sapaty P.S.: Managing distributed systems with spatial grasp patterns. Math. Mach. Syst., (4), (2023)
175. Sapaty, P.S.: Basic network operations under spatial grasp model. Sci Set J Physics., **3**(1), 01–16 (2024). https://mkscienceset.com/articles_file/500-_article1708322845.pdf
176. Sapaty, P.S.: Spatial networks as models for organoid cultures and brain research. Math. Mach. Syst., (1), (2024)
177. Bondarenko, A.T., Mikhalevich, S.B., Nikitin, A.I., Sapaty, P.S.: Software of BESM-6 computer for communication with peripheral computers via telephone channels, in Computer Software, vol. 5. Inst. of Cybernetics Press, Kiev (1970)
178. Sapaty, P.S.: A Method of organization of an intercomputer dialogue in the radial computer systems, in the design of software and hardware for automatic control systems. Inst. of Cybernetics Press, Kiev (1973)
179. Sapaty, S.: A wave language for parallel processing of semantic networks. Comput. Artif. Intell. **5**(4), 289–314 (1986)

180. Sapaty, P.S.: The WAVE-0 language as a framework of navigational structures for knowledge bases using semantic networks. Proc USSR Acad. Sciences. Tech. Cybern., (5), (1986)
181. Sapaty, P.S.: The WAVE-1: A new ideology and language of distributed processing on graphs and networks. Comput. Artif. Intell., (5), (1987)
182. Sapaty, P.S.: The WAVE model for advanced knowledge processing. Report No. OUEL 1803/89. University of Oxford, England, (1989)
183. Sapaty, P.S.: The WAVE model for advanced knowledge processing. In: Ambler, A.P., Agrawal, P., Moore, W.R., (eds.) CAD accelerators. Elsevier Science Publ. B.V., (1990)
184. Sapaty, P.S.: The WAVE machine project, pp. 28–30. Proc. IFIP Workshop on Silicon Architectures for Neural Nets, St. Paul de Vence, France, November (1990)
185. Sapaty, P.S.: The WAVE paradigm. Report No. 17/92. Faculty of Informatics, University of Karlsruhe, (1992)
186. Sapaty, P.S.: The WAVE paradigm. In: Proc JICSLP'92 post-conference joint workshop on distributed and parallel implementations of logic programming systems, Washington, D.C., (1992)
187. Sapaty, P.S.: Distributed modeling of cooperative behavior by mobile agents. In: Proc sixth conference on computer generated forces and behavioral representation, IST UCF, Orlando, FL, pp. 561–599 (1996)

The manufacturer's authorised representative in the EU is Springer Nature Customer Service Centre GmbH, Europaplatz 3, 69115 Heidelberg, Germany. If you have any concerns regarding our products, please contact ProductSafety@springernature.com

Printed and bound by CPI Group (UK) Ltd, Croydon, CR0 4YY
01/10/2026
02225970-0004